I0474670

A STUDY OF THE SPECIAL THEORY OF RELATIVITY

Richard Banner

Copyright © 2019 by Richard Banner

All rights reserved. No part of this publication may be reproduced, stored in a retrieval system, or transmitted in any form or by any means, electronic, mechanical, photocopying, recording or otherwise, without the prior permission of the publishers.

PREFACE

The Special Theory of Relativity is a very important piece of work in physics. It is a work of recognised great genius which has made its creator Albert Einstein famous as well as the icon for genius.

This book is a survey of the famous theory with alternative views and suggestions.

Richard Banner, PhD

CONTENTS

1 ANOMALY IN COMPUTING SPEED OF LIGHT

The Special Theory of Relativity postulates that the speed of light always remains constant at 186,000 miles per second at all inertial frames. This chapter describes an anomaly in the standard computation pertaining to the constancy of the speed of light.

The Special Theory of Relativity posits that a person on a moving vehicle, e.g., a very fast moving train (moving frame), traveling at close to the speed of a beam of light (moving frame) in the same direction would find the speed of the beam of light (moving frame) to be unchanged at 186,000 miles per second, instead of the difference between the speed of the very fast moving train (moving frame) and the speed of the beam of light (moving frame), which would normally be the case. This is because, according to the Special Theory, on the very fast moving train (moving frame) approaching the speed of light the clock therein used to gauge the time traveled by the beam of light (moving frame) has slowed down by the same degree (say X %) as the ruler or measuring device (stated as meter stick or measuring rod in some texts) therein used to gauge the distance traveled by the beam of light (moving frame) has contracted in length in the direction of the very fast moving train's motion (also X %), the greater the very fast moving train's traveling speed the more the clock slows down and the greater the length contraction of the ruler or measuring device. This is expressed in the following equation (the speed of the beam of light (moving frame) being the distance it traveled divided by the time it took to travel this distance), which is in accordance with the Special Theory of Relativity:-

(186,000 miles - X % of 186,000 miles) ÷ (1 second - X % of 1 second) = 186,000 miles per second

In other words, there has to be a same percentage decrease in the time gauged and the distance gauged due to the respective slowing down of the clock and contracting in length of the ruler or measuring device therein the very fast moving train (moving frame), in order for the speed of the beam of light (moving frame) to remain constant, which is consistent with mathematical logic - this condition must indeed apply in order for the speed of the beam of light (moving frame) to remain constant.

There is however something not quite usual related to the above concept. According to the Special Theory of Relativity, the person on the very fast moving train traveling at close to the speed of light (moving frame) gauging the speed of the beam of light traveling in the same direction (moving frame) would not notice that the clock on his very fast moving train (moving frame) has slowed down and the ruler or measuring device therein has contracted in length in the direction of motion. In other words, everything would appear normal to him, despite the fact that his clock has actually slowed down and his ruler or measuring device has actually contracted in length in the direction of motion, as is postulated by the Special Theory of Relativity. But, according to the Special Theory of Relativity, when he (moving frame) compares himself to a person on the ground who is not moving (stationary frame), he could even consider himself stationary (stationary frame) while thinking that the person on the ground (who is not moving) is actually moving (moving frame), i.e., all movements are relative. He (moving frame) would notice that the clock on the ground (stationary frame) is slower and the ruler or measuring device on the ground (stationary frame) is shorter. The person on the ground who is not moving (stationary frame) would also notice that the clock on the very fast moving train (moving frame) is slower, the ruler or measuring device therein is shorter, and, the length of the very fast moving train (moving frame) is shorter. In other words, both the train-traveler (moving frame) and the person on the ground who is not moving (stationary frame) would notice that the other's clock is slower and the other's ruler or measuring device is shorter, and, according to the Special Theory of Relativity, the slowing down of clocks and the shortening of rulers or measuring devices would appear to be by the same degree (X %) for both.

But, it is actually the clock on the very fast moving train traveling at close to the speed of the beam of light in the same direction (moving frame) which has slowed down and the ruler or measuring device therein which has contracted in length (in the direction of motion) as is postulated by the Special Theory of Relativity, and not those on the ground (stationary frame). To the traveler on the very fast moving train (moving frame) who is gauging the speed of the beam of light traveling in the same direction (moving frame), the beam of light (moving frame) appears to take less time (time dilation) to travel a shorter distance (length contraction), which, according to the Special Theory of Relativity and in accordance with the following equation, explains the constancy of the speed of light at all inertial frames:-

(186,000 miles - X % of 186,000 miles) ÷ (1 second - X % of 1 second) = 186,000 miles per second

The speed of the beam of light (moving frame) is obtained by dividing the distance traveled by the beam of light (as gauged by the ruler or measuring device on the very fast moving train traveling at close to the speed of light - moving frame) by the time it took to travel that distance by the beam of light (as gauged by the clock therein the very fast moving train - moving frame), the gauging being carried out by the traveler on the very fast moving train (moving frame). In the above example, the clock on the very fast moving train traveling at close to the speed of light (moving frame) slows down and gauges a slower time, X % slower. The ruler or measuring device therein also contracts in length in the direction of motion by X %; however because of this it should gauge any object as "longer" due to a change in the scale of the ruler or measuring device on contracting in length in the direction of motion (see Appendix 1), and, this evidently gives rise to an anomaly when computing the speed of the beam of light (moving frame). This anomaly in the computation of the speed of light is described in the example below.

The person on the very fast moving train (moving frame) traveling at close to the speed of the beam of light (moving frame) in the same direction could be considered one inertial frame, and, the ground level (stationary frame) could be considered another inertial frame. If, traveling at close to the speed of light, both his clock has slowed down and his ruler or measuring device has contracted in length in the direction of motion by the same degree or percentage, say 50 %, the beam of light (moving frame) whose speed he is gauging, according to the Special Theory of Relativity, would also appear to him to have traveled a distance (from one designated point to another designated point - reference, stationary frame) which is shortened by 50 %, i.e., the distance between the two designated points (reference, stationary frame) appears to him to have shortened by 50 %.

Let us look at the following self-explanatory diagrammatic example, which contains the above-said anomaly:-

Gauging Of The Speed Of A Beam Of Light At Close To The Speed Of Light When, For Example, The Clock Has Slowed Down By 50 % And The Ruler Has Contracted In Length In The Direction Of Motion By 50 %
How do we gauge the speed of a beam of light (moving frame) while traveling at close to the speed of light on a very fast moving train (moving frame) in the same direction? We do so as follows:-

a) The beam of light (moving frame) travels from one point to another point, say, from Point A to Point B (the distance between Point A and Point B being a reference, stationary frame next to the beam of light, say, on the embankment next to the railway tracks).

b) Let us say that the distance from Point A to Point B (reference, stationary frame) is 1 metre and the beam of light (moving frame) takes x second to travel this 1 metre from Point A to Point B (reference, stationary frame), i.e., the speed of the beam of light (moving frame) is 1 metre per x second. In other words, as gauged from the ground level (stationary frame) the speed of the beam of light (moving frame) traveling from Point A to Point B (reference, stationary frame) is 1 metre per x second, which is depicted as follows:

```
beam of light travels 1 metre from Point A to Point B (speed is 1 metre
------------------------------------------------------------> per x second)
l------------------------------------------------------------l x second (before clock
<------------------------ 1 metre ----------------------> slows down 50 %)
    (length of ruler before 50 % length contraction)
```

```
_____
l                                                 l (ruler before 50 %
l_____l length contraction)
            1-metre ruler
```

c) Next, the beam of light (moving frame) is gauged from a very fast moving train (moving frame) traveling in the same direction as the beam of light (moving frame), at close to the speed of light, when, say, the clock used to time the distance traveled by the beam of light (moving frame) from Point A to Point B (reference, stationary frame) on the embankment besides the railway tracks has slowed down by 50 %, the 1-metre-long ruler used to measure the distance traveled has contracted in length in the direction of motion by 50 %, and the distance traveled by the beam of light (moving frame) from Point A to Point B (reference, stationary frame) has also shortened by the same degree, 50 % (in the eyes of the train-traveler (moving frame) gauging the speed of the beam of light (moving frame), which is in accordance with the Special Theory of Relativity) - i.e., the distance between Point A & Point B (reference, stationary frame) on the embankment besides the railway tracks has shortened from 1 metre by 50 % to 0.5 metre as seen by the train-traveler (moving frame) who is gauging the speed of the beam of light (moving frame), which is in accordance with the Special Theory of Relativity. Note that, in accordance with the Special Theory of Relativity, the train-traveler (moving frame) does not notice that his clock has slowed down by 50 % and his 1-metre-long ruler

has contracted in length in the direction of motion by 50 % while his train
is moving at close to the speed of light. All this is depicted as follows:

beam of light (contracts in length by 50 % in the eyes of the train-traveler - speed is now 0.5
metre per **1/8** x second, or, 4 metres per x second)
---------------->

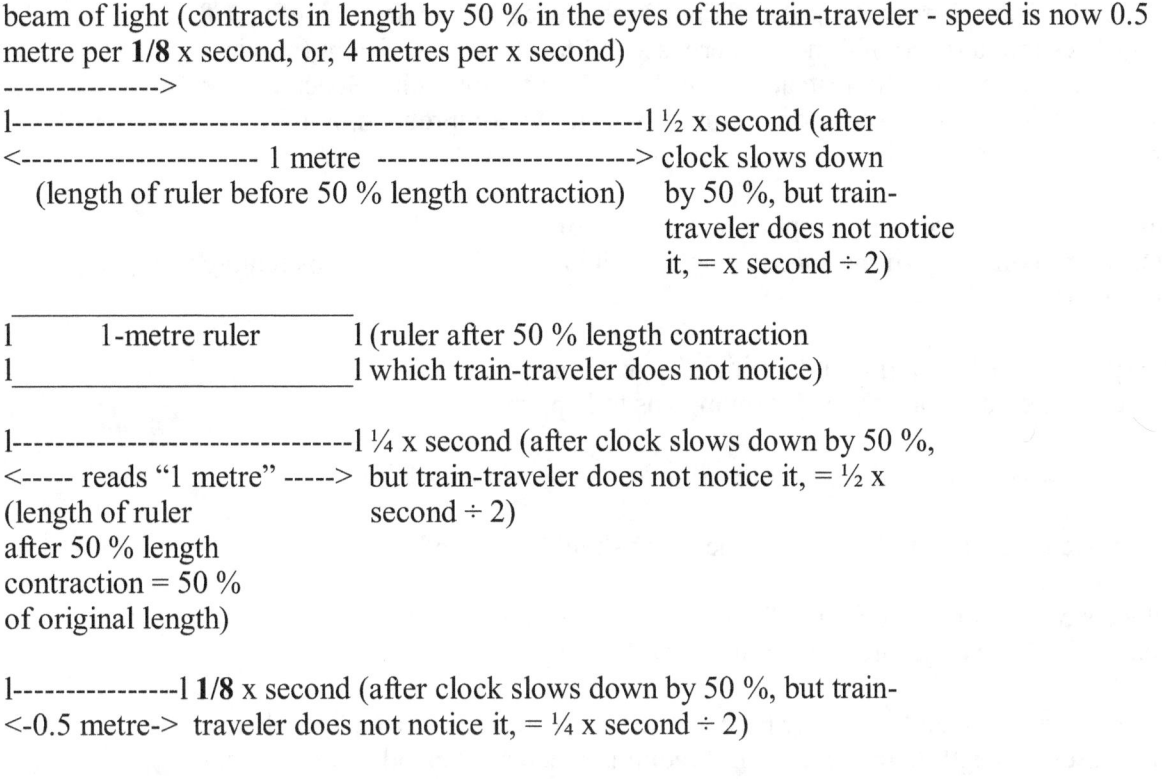

l--l ½ x second (after
<----------------------- 1 metre ------------------------> clock slows down
    (length of ruler before 50 % length contraction)   by 50 %, but train-
                                                      traveler does not notice
                                                      it, = x second ÷ 2)

l   1-metre ruler   l (ruler after 50 % length contraction
l_____l which train-traveler does not notice)

l--------------------------------l ¼ x second (after clock slows down by 50 %,
<----- reads "1 metre" -----> but train-traveler does not notice it, = ½ x
(length of ruler   second ÷ 2)
after 50 % length
contraction = 50 %
of original length)

l----------------l **1/8** x second (after clock slows down by 50 %, but train-
<-0.5 metre-> traveler does not notice it, = ¼ x second ÷ 2)

(train-traveler sees this 0.5 metre as the distance from Point A to Point B traveled by the
beam of light after the distance has shortened by 50 % in accordance with the Special Theory
of Relativity)

The train-traveler (moving frame) obtains the speed of the beam of light (moving frame) by
dividing the distance between Point A & Point B (reference, stationary frame), which in his
eyes, in accordance with the Special Theory of Relativity, has contracted by 50 % from 1
metre to 0.5 metre, by the time the beam of light (moving frame) takes to travel this distance
(0.5 metre).

As is evident from the above diagrams, the train-traveler's clock, which has slowed down by 50 %, would time the distance (0.5 metre now) traveled by the beam of light (moving frame) 50 % slower of course; it would gauge the time the beam of light (moving frame) takes to travel the distance of 0.5 metre from Point A to Point B (reference, stationary frame) as **1/8** x second. This translates into a speed of 4 metres per x second for the beam of light (moving frame), i.e., 4 times the speed of light, instead of just the speed of light which Special Relativity postulates. (The above description might be difficult to comprehend. For an alternative description, see Appendix 2.)

This is an anomaly and it contradicts an important postulate of the Special Theory of Relativity, namely, the constancy of the speed of light at all inertial frames. All this requires attention and correction.

In order for the speed of the beam of light to remain/appear constant in this instance, i.e., remain at 1 metre per x second, one of the following has to happen:-

1) When the clock slows down by 50 %, the ruler should increase in length by 100 %.
2) When the ruler decreases in length by 50 %, the clock should quicken by 100 %.
3) When the clock slows down by 50 % and the ruler decreases in length by 50 %, the beam of light (moving frame) should slow down by 400 %.

As a matter of fact, depending on how many percent the train-traveler's clock slows down and his ruler decreases in length in the direction of motion, which all depends on how fast the train is traveling, the speed of the beam of light would appear variable, and not constant, to the train-traveler. See the two other examples in Appendix 3.

APPENDIX 1
For example, we have an atlas of size X feet in length by Y feet in breadth with a scale of 1 is to 100,000 (1 inch on the atlas represents 100,000 inches on the actual ground) which shows only a portion of our globe. To have the whole globe represented in this atlas X feet in length by Y feet in breadth, it is evident that we have to change its scale, e.g., change its scale to 1 is to 1,000,000 (1 inch on the atlas represents 1,000,000 inches on the actual ground). The above-mentioned ruler or measuring device which has contracted in length is analogous to an

atlas whose scale has been changed to allow it to represent a larger area (i.e., the whole globe), e.g., from 1 is to 100,000 to 1 is to 1,000,000 - with the new scale of 1 is to 1,000,000, 1 inch on the atlas now represents a length of 1,000,000 inches instead of 100,000 inches on the actual ground. If we change the scale of the above-mentioned atlas which shows only a portion of our globe to 1 is to 1,000,000 from 1 is to 100,000, we would have a much smaller atlas showing the same portion of our globe, whose dimensions would be 1/10 X feet in length by 1/10 Y feet in breadth, a contracted atlas which is analogous to the above-mentioned ruler or measuring device which has contracted in length. A 1-metre-long ruler which has contracted in length by 50 %, e.g., would now gauge the length of 1 metre as 2 metres (and not 0.5 metre in accordance with the Special Theory of Relativity), and this is evidently the cause of the above-described anomaly relating to the computation of the speed of light. In fact, for the 1-metre-long ruler to gauge the length of 1 metre as 0.5 metre it would have to increase in length by 100 %. A shortened ruler would gauge an object as "longer" while a lengthened ruler would gauge an object as "shorter".

Let us look at a simple example here. Say, the 1-metre-long ruler used to gauge distance has contracted in length in the direction of motion by 20 %. The clock used to gauge time, which has also slowed down by 20 %, according to the Special Theory of Relativity, would now gauge the time taken, say X, to travel the distance between two designated points (reference, stationary frame), say Y, as having decreased by 20 % to become 0.8 X. Though the 1-metre-long ruler, which has contracted in length by 20 %, still reads "1 metre" in length, it is in effect shorter by 20 % (actually only 0.8 metre in length). Therefore, when it gauges the distance traveled, Y, above, this distance Y would now be gauged as 1.25 Y, and not 0.8 Y in accordance with the Special Theory of Relativity. As stated above, the Special Theory of Relativity theorizes that for a beam of light (moving frame) to remain constant in speed the beam of light (moving frame) has to take less time (time dilation) to travel a shorter distance (length contraction) - in effect, X % less time to travel a distance shorter by X %, in accordance with the following equation, which implies that the speed of the beam of light (moving frame) would remain constant, e.g., 0.8 X (time) to travel 0.8 Y (distance) after "time dilation" and "length contraction":-

(186,000 miles - X % of 186,000 miles) ÷ (1 second - X % of 1 second) = 186,000 miles per second

But, as explained above, this would not be the case; the beam of light (moving frame) would have been gauged as having taken 0.8 X (time) to travel 1.25 Y (distance). This is an anomaly in the Special Theory of Relativity.

APPENDIX 2

<u>Alternative Description - Gauging Of The Speed Of A Beam Of Light At Close To The
Speed Of Light When, For Example, The Clock Has Slowed Down By 50 % And The Ruler
Has Contracted In Length In The Direction Of Motion By 50 %</u>
The description below is based on the following conditions ((i), (ii), (iii), (iv) & (v)) which
are related to the Special Theory of Relativity:-

i) Clocks slow down and rulers contract in length in the direction of motion, at close to the
speed of light, but these would not be noticed by the person traveling at close to the
speed of light (moving frame) who possess such clocks and rulers. These do not apply to
the clock and ruler of the person on the ground who is not moving (stationary frame).

ii) The train-traveler traveling at close to the speed of light (moving frame) and the person
on the ground who is not moving (stationary frame) would notice one another's clock
and ruler as respectively slower and shorter than his own. Both of them would each
notice the other's clock slowing down and ruler length shortening to the same degree
(X % for the clock and ruler of the train-traveler traveling at close to the speed of light
(moving frame) and also X % for the clock and ruler of the person on the ground who is
not moving (stationary frame)). Though the train-traveler's clock has slowed down and
his ruler has contracted in length in the direction of motion while he is traveling at
close to the speed of light, as is described in (i) above, he would not notice these
happenings. All this does not apply to the clock and ruler (which are not affected by (i)
above) of the person on the ground who is not moving (stationary frame).

iii) All movements are relative. When the train-traveler traveling at close to the speed of
light (moving frame) compares himself to a person on the ground who is not moving
(stationary frame), he could consider himself stationary (stationary frame) while
thinking that the person on the ground (who is not moving) is actually moving (moving
frame) - both parties could each regard themselves as stationary (stationary frame) and
consider the other party in motion (moving frame).

iv) The train-traveler traveling at close to the speed of light (moving frame) would see the
distance from Point A to Point B at the embankment besides the railway tracks
(reference, stationary frame - you might substitute this distance with a ruler) as having
shortened. This shortening (by X %) of the distance from Point A to point B (reference,
stationary frame) would be by the same degree as his own clock has slowed down
(X %) and his own ruler has contracted in length in the direction of motion (X %), both
of which he does not notice. This point is the equivalent of Point (ii) above.

v) The speed of the beam of light (moving frame) is obtained by dividing the distance traveled by the beam of light (as gauged by the ruler on the very fast moving train traveling at close to the speed of light - moving frame) by the time it took to travel that distance by the beam of light (as gauged by the clock therein the very fast moving train - moving frame), the gauging being carried out by the traveler on the very fast moving train (moving frame), the greater the very fast moving train's traveling speed the more the clock slows down and the greater the length contraction of the ruler. For the speed of the beam of light (moving frame) to be measured as constant, the clock in the train traveling at close to the speed of light (moving frame) used to gauge the time traveled by the beam of light (moving frame) has to gauge a time which has slowed down by the same degree (X %) as the distance traveled by the beam of light (moving frame) which has shortened (also X %) as is gauged by the ruler therein.

1) Before the distance from Point A to Point B (reference, stationary frame) has shortened by 50 % (as is gauged at the ground level - stationary frame):-

   a) Distance from Point A to Point B (reference, stationary frame) = 1 metre
   b) Time taken by the beam of light (moving frame) to travel this 1 metre = x second,
      i.e., speed of the beam of light (moving frame) = <u>1 metre per x second</u>

The above distance of 1 metre in (a) is gauged at the ground level (stationary frame) with a ruler which has not contracted in length (by 50 %, as yet). The above time of x second in (b) is gauged at the ground level (stationary frame) with a clock which has not slowed down (by 50 %, as yet).

2) a) After the ruler on the moving train (moving frame) has contracted in length in the direction of motion by 50 %, it would gauge the above-mentioned 1 metre (distance from Point A to Point B (reference, stationary frame), before shortening) as <u>2 metres</u>. (A shortened ruler gauges an object as "longer" while a lengthened ruler gauges an object as "shorter", due to the different scales, as is explained in Appendix 1.)
   b) After the clock on the moving train (moving frame) has slowed down by 50 %, at the same time that the ruler has contracted in length by 50 %, it would gauge the above-mentioned x second taken to travel this measured distance of 2 metres as <u>1/2 x second</u>.
   c) That is, the beam of light (moving frame) is now gauged from the moving train (moving frame) as requiring 1/2 x second to travel the distance of 2 metres (before the shortening of the distance between Point A & Point B (we assume here that the distance between Point A & Point B has not shortened) - in accordance with the Special Theory of Relativity this distance would also be shortened (also by 50 %) as

seen by the train-traveler, as is described below); this translates into a speed of 4 metres per x second, or, 4 X speed of light.

3) After the distance from Point A to Point B (reference, stationary frame) has shortened by 50 %, as is seen by the train-traveler traveling at close to the speed of light (moving frame), in accordance with the Special Theory of Relativity:-

   a) The 1-metre ruler which has contracted in length in the direction of motion by 50 % would gauge the above-mentioned 50 % shortened distance from Point A to Point B (reference, stationary frame) as 50 % shorter than before (i.e., 0.5 metre instead of 1 metre), and would read "0.5 metre" (instead of "1 metre" - full length of the 1-metre ruler) on its 50 % shortened length, which is in accordance with the Special Theory of Relativity. (The ruler now has a different scale as compared to its scale before length contraction. See Appendix 1 for explanation.)
   b) When the scale of the ruler has changed due to the ruler's length contraction (by 50 %), and the clock has slowed down (by 50 %), the beam of light (moving frame) being gauged as requiring 1/2 x second to travel the distance of 2 metres, as is described in (2) above, the time taken by the beam of light (moving frame) to travel 0.5 metre now = (0.5 metre ÷ 2 metres) x 1/2 x second = 1/8 x second, i.e., the speed of the beam of light (moving frame) is now 0.5 metre per 1/8 x second, or, 4 metres per x second/4 X speed of light, which is an anomaly (the speed of the beam of light should have remained 1 metre per x second/1 X speed of light, as is postulated by the Special Theory of Relativity).

The above computations have been carried out in accordance with the conditions which are related to the Special Theory of Relativity. The said anomaly is evidently due to the change of the scale of the ruler which has contracted in length in the direction of motion by 50 %.

APPENDIX 3
Two Other Examples - Gauging Of The Speed Of A Beam Of Light At Close To The Speed Of Light When The Clock Has Slowed Down And The Ruler Has Contracted In Length In The Direction Of Motion
The descriptions of the two examples below are based on the following conditions ((i), (ii), (iii), (iv) & (v)) which are related to the Special Theory of Relativity:-

i) Clocks slow down and rulers contract in length in the direction of motion, at close to the speed of light, but these would not be noticed by the person traveling at close to the speed of light (moving frame) who possess such clocks and rulers. These do not apply to the clock and ruler of the person on the ground who is not moving (stationary frame).

ii) The train-traveler traveling at close to the speed of light (moving frame) and the person on the ground who is not moving (stationary frame) would notice one another's clock and ruler as respectively slower and shorter than his own. Both of them would each notice the other's clock slowing down and ruler length shortening to the same degree (X % for the clock and ruler of the train-traveler traveling at close to the speed of light (moving frame) and also X % for the clock and ruler of the person on the ground who is not moving (stationary frame)). Though the train-traveler's clock has slowed down and his ruler has contracted in length in the direction of motion while he is traveling at close to the speed of light, as is described in (i) above, he would not notice these happenings. All this does not apply to the clock and ruler (which are not affected by (i) above) of the person on the ground who is not moving (stationary frame).

iii) All movements are relative. When the train-traveler traveling at close to the speed of light (moving frame) compares himself to a person on the ground who is not moving (stationary frame), he could consider himself stationary (stationary frame) while thinking that the person on the ground (who is not moving) is actually moving (moving frame) - both parties could each regard themselves as stationary (stationary frame) and consider the other party in motion (moving frame).

iv) The train-traveler traveling at close to the speed of light (moving frame) would see the distance from Point A to Point B at the embankment besides the railway tracks (reference, stationary frame - you might substitute this distance with a ruler) as having shortened. This shortening (by X %) of the distance from Point A to point B (reference, stationary frame) would be by the same degree as his own clock has slowed down (X %) and his own ruler has contracted in length in the direction of motion (X %), both of which he does not notice. This point is the equivalent of Point (ii) above.

v) The speed of the beam of light (moving frame) is obtained by dividing the distance traveled by the beam of light (as gauged by the ruler on the very fast moving train traveling at close to the speed of light - moving frame) by the time it took to travel that distance by the beam of light (as gauged by the clock therein the very fast moving train - moving frame), the gauging being carried out by the traveler on the very fast moving train (moving frame), the greater the very fast moving train's traveling speed the more the clock slows down and the greater the length contraction of the ruler. For the speed of the beam of light (moving frame) to be measured as constant, the clock in the train traveling at close to the speed of light (moving frame) used to gauge the time traveled by the beam of light (moving frame) has to gauge a time which has slowed down by the same degree (X %) as the distance traveled by the beam of light (moving frame) which has shortened (also X %) as is gauged by the ruler therein.

A) Example 1 - Gauging Of The Speed Of A Beam Of Light At Close To The Speed Of Light When The Clock Has Slowed Down By 25 % And The Ruler Has Contracted In Length In The Direction Of Motion By 25 %

1) Before the distance from Point A to Point B (reference, stationary frame) has shortened by 25 % (as is gauged at the ground level - stationary frame):-

   a) Distance from Point A to Point B (reference, stationary frame) = 1 metre
   b) Time taken by the beam of light (moving frame) to travel this 1 metre = x second,
      i.e., speed of the beam of light (moving frame) = 1 metre per x second

   The above distance of 1 metre in (a) is gauged at the ground level (stationary frame) with a ruler which has not contracted in length (by 25 %, as yet). The above time of x second in (b) is gauged at the ground level (stationary frame) with a clock which has not slowed down (by 25 %, as yet).

2) a) After the ruler on the moving train (moving frame) has contracted in length in the direction of motion by 25 %, it would gauge the above-mentioned 1 metre (distance from Point A to Point B (reference, stationary frame), before shortening) as 1.33 metres. (A shortened ruler gauges an object as "longer" while a lengthened ruler gauges an object as "shorter", due to the different scales, as is explained in Appendix 1.)
   b) After the clock on the moving train (moving frame) has slowed down by 25 %, at the same time that the ruler has contracted in length by 25 %, it would gauge the above-mentioned x second taken to travel this measured distance of 1.33 metres as 3/4 x second.
   c) That is, the beam of light (moving frame) is now gauged from the moving train (moving frame) as requiring 3/4 x second to travel the distance of 1.33 metres (before the shortening of the distance between Point A & Point B (we assume here that the distance between Point A & Point B has not shortened) - in accordance with the Special Theory of Relativity this distance would also be shortened (also by 25 %) as seen by the train-traveler, as is described below); this translates into a speed of 1.77 metres per x second, or, 1.77 X speed of light.

3) After the distance from Point A to Point B (reference, stationary frame) has shortened by 25 %, as is seen by the train-traveler traveling at close to the speed of light (moving

frame), in accordance with the Special Theory of Relativity:-

a) The 1-metre ruler which has contracted in length in the direction of motion by 25 % would gauge the above-mentioned 25 % shortened distance from Point A to Point B (reference, stationary frame) as 25 % shorter than before (i.e., 0.75 metre instead of 1 metre), and would read "0.75 metre" (instead of "1 metre" - full length of the 1-metre ruler) on its 25 % shortened length, which is in accordance with the Special Theory of Relativity. (The ruler now has a different scale as compared to its scale before length contraction. See Appendix 1 for explanation.)

b) When the scale of the ruler has changed due to the ruler's length contraction (by 25 %), and the clock has slowed down (by 25 %), the beam of light (moving frame) being gauged as requiring 3/4 x second to travel the distance of 1.33 metres, as is described in (2) above, the time taken by the beam of light (moving frame) to travel 0.75 metre now = (0.75 metre ÷ 1.33 metres) x 3/4 x second = 0.423 x second, i.e., the speed of the beam of light (moving frame) is now 0.75 metre per 0.423 x second, or, 1.77 metres per x second/1.77 X speed of light, which is an anomaly (the speed of the beam of light should have remained 1 metre per x second/1 X speed of light, as is postulated by the Special Theory of Relativity).

The above computations have been carried out in accordance with the conditions which are related to the Special Theory of Relativity. The said anomaly is evidently due to the change of the scale of the ruler which has contracted in length in the direction of motion by 25 %.

In order for the speed of the beam of light to remain/appear constant in this instance, i.e., remain at 1 metre per x second, one of the following has to happen:-

1) When the clock slows down by 25 %, the ruler should increase in length by 33 %.

2) When the ruler decreases in length by 25 %, the clock should quicken by 33 %.

3) When the clock slows down by 25 % and the ruler decreases in length by 25 %, the beam of light (moving frame) should slow down by 177 %.

B) Example 2 - Gauging Of The Speed Of A Beam Of Light At Close To The Speed Of Light When The Clock Has Slowed Down By 90 % And The Ruler Has Contracted In Length In The Direction Of Motion By 90 %

1) Before the distance from Point A to Point B (reference, stationary frame) has shortened by 90 % (as is gauged at the ground level - stationary frame):-

a) Distance from Point A to Point B (reference, stationary frame) = 1 metre

b) Time taken by the beam of light (moving frame) to travel this 1 metre = x second, i.e., speed of the beam of light (moving frame) = 1 metre per x second

The above distance of 1 metre in (a) is gauged at the ground level (stationary frame) with a ruler which has not contracted in length (by 90 %, as yet). The above time of x second in (b) is gauged at the ground level (stationary frame) with a clock which has not slowed down (by 90 %, as yet).

2) a) After the ruler on the moving train (moving frame) has contracted in length in the direction of motion by 90 %, it would gauge the above-mentioned 1 metre (distance from Point A to Point B (reference, stationary frame), before shortening) as 10 metres. (A shortened ruler gauges an object as "longer" while a lengthened ruler gauges an object as "shorter", due to the different scales, as is explained in Appendix 1.)

   b) After the clock on the moving train (moving frame) has slowed down by 90 %, at the same time that the ruler has contracted in length by 90 %, it would gauge the above-mentioned x second taken to travel this measured distance of 10 metres as 1/10 x second.

   c) That is, the beam of light (moving frame) is now gauged from the moving train (moving frame) as requiring 1/10 x second to travel the distance of 10 metres (before the shortening of the distance between Point A & Point B (we assume here that the distance between Point A & Point B has not shortened) - in accordance with the Special Theory of Relativity this distance would also be shortened (also by 90 %) as seen by the train-traveler, as is described below); this translates into a speed of 100 metres per x second, or, 100 X speed of light.

3) After the distance from Point A to Point B (reference, stationary frame) has shortened by 90 %, as is seen by the train-traveler traveling at close to the speed of light (moving frame), in accordance with the Special Theory of Relativity:-

   a) The 1-metre ruler which has contracted in length in the direction of motion by 90 % would gauge the above-mentioned 90 % shortened distance from Point A to Point B (reference, stationary frame) as 90 % shorter than before (i.e., 0.1 metre instead of 1 metre), and would read "0.1 metre" (instead of "1 metre" - full length of the 1-metre ruler) on its 90 % shortened length, which is in accordance with the Special Theory of Relativity. (The ruler now has a different scale as compared to its scale before length contraction. See Appendix 1 for explanation.)

   b) When the scale of the ruler has changed due to the ruler's length contraction (by 90 %), and the clock has slowed down (by 90 %), the beam of light (moving frame) being gauged as requiring 1/10 x second to travel the distance of 10 metres, as is described in (2) above, the time taken by the beam of light (moving frame) to travel 0.1 metre now = (0.1 metre ÷ 10 metres) x 1/10 x second = 0.001 x second, i.e., the speed of the beam of light (moving frame) is now 0.1 metre per 0.001 x second, or, 100 metres per x second/100 X speed of light, which is an anomaly (the speed of the beam of light should have remained 1 metre per x second/1 X speed of light, as is postulated by the Special Theory of Relativity).

The above computations have been carried out in accordance with the conditions which are related to the Special Theory of Relativity. The said anomaly is evidently due to the change of the scale of the ruler which has contracted in length in the direction of motion by 90 %.

In order for the speed of the beam of light to remain/appear constant in this instance, i.e., remain at 1 metre per x second, one of the following has to happen:-

1) When the clock slows down by 90 %, the ruler should increase in length by 900 %.
2) When the ruler decreases in length by 90 %, the clock should quicken by 900 %.
3) When the clock slows down by 90 % and the ruler decreases in length by 90 %, the beam of light (moving frame) should slow down by 10,000 %.

It is therefore evident that the speed of the beam of light could not be constant and is variable, which implies that there is some logical error in the Special Theory of Relativity. For instance, quantum particles are able to teleport or transport themselves to another location instantaneously, which is an example of faster than light travel.

2 SUBTLETY AND PHILOSOPHICAL ASPECT OF THEORY

According to Einstein's Special Theory of Relativity, the speed of light always remains constant at 186,000 miles per second regardless of whether it is gauged from a stationary reference point, a moving reference point or any other reference points, no object could travel faster than 186,000 miles per second (the speed of light itself) because the mass of the object would then be so great (infinitely great) that it could not accelerate anymore, on approaching the speed of light a moving object contracts in length in the direction of motion while a clock gauging the time slows down, at the speed of light the length of the moving object contracts to zero while the clock (and time) becomes at a standstill, and, importantly, the mass of an object multiplied by the square of the speed of light gives energy ($E = MC^2$), i.e., mass could be converted to energy and vice versa; on approaching the speed of light the brain and bodily functions of a person slow down; observers do not agree on the simultaneity of events - two simultaneous events for one observer might not be simultaneous for another; in the Special Theory time-travel (in the space-time continuum) is an apparent possibility. A deeper look at the Special Theory of Relativity is presented in this chapter, employing some strong, subtle and important mathematical reasoning in the process.

The Special Theory of Relativity posits that on approaching the speed of light clocks slow down, moving objects contract in length in the direction of motion, and, a person's brain and bodily functions slow down. For example, a person on a moving vehicle (moving frame), which travels besides a beam of light (moving frame) in the same direction and almost as fast as the beam of light (moving frame) itself, gauges the speed of the beam of light (moving frame). The Special Theory postulates that this person on the moving vehicle (moving frame) traveling at almost the speed of the beam of light (moving frame) would find the speed of the beam of light (moving frame) to be unchanged at 186,000 miles per second, instead of the difference between the speed of the moving vehicle (moving frame) and the speed of the beam of light (moving frame), which would normally be the case; this is because, according to the Special Theory, on the moving vehicle (moving frame) approaching the speed of light the clock therein used to gauge the time traveled by the beam of light (moving frame) has slowed down by the same degree (say X %) as the ruler or measuring device therein used to gauge the distance traveled by the beam of light (moving frame) has contracted in length in the direction of the vehicle's motion (also X %), the greater the moving vehicle's traveling speed the more the clock slows down and the greater the length contraction of the ruler or measuring device.

This is expressed in the following equation, which is in accordance with the Special Theory of Relativity:-

(186,000 miles - X % of 186,000 miles) ÷ (1 second - X % of 1 second) = 186,000 miles per second

A person traveling at almost the speed of light (as in the case of the person in the above-mentioned vehicle traveling at almost the speed of light (moving frame) besides a beam of light (moving frame)) would experience the slowing down of time and age more slowly (without being aware of it), as stipulated by the Special Theory of Relativity. This "time dilation" effect is described by the following equation:-

$$t = \frac{t'}{\sqrt{1 - v^2/c^2}}$$

t = time gauged by a clock at the ground level (stationary frame); t' = time gauged by a clock on the above-mentioned vehicle traveling at almost the speed of light (moving frame); c = speed of light = 186,000 miles per second; v = speed of the above-mentioned vehicle traveling at almost the speed of light (moving frame), e.g., 0.9c; t' would be a fraction of t, i.e., t' < t, e.g., if t' = 24 years, v = 0.8c, then t = 40 years.

But (and this is a very important "but"), if the moving vehicle's clock had not slowed down and its ruler or measuring device had not contracted in length (i.e., under normal conditions), the speed of the light beam (moving frame), as gauged from the vehicle traveling besides it at almost the speed of light (moving frame), would have had been the difference between the speed of the light beam (moving frame) and the speed of the moving vehicle (moving frame), e.g., 186,000 miles per second (speed of the light beam) minus 185,990 miles per second (speed of the moving vehicle), which is equal to 10 miles per second.

Thus, the speed of the light beam (moving frame), i.e., 186,000 miles per second, as gauged from the vehicle traveling besides it at almost the same speed (moving frame) in the same direction is evidently an *illusion*, which is the result of the slowing down of the clock, and, the brain and bodily functions of the person, on the moving vehicle (moving frame), and the contraction in the length of the ruler or measuring device (in the direction of motion) therein, as stipulated by the Special Theory of Relativity. This is somewhat similar to the situation whereby a driver in a car which is actually cruising at 60 miles per hour believes that his car is traveling at 30 miles per hour because the car's speedometer, which happens to be faulty, gives a reading of the car's cruising speed as 30 miles per hour instead of 60 miles per hour, i.e., the driver is misled by the car's faulty speedometer.

In the above-mentioned case, the person on the vehicle traveling at almost the speed of light (moving frame), say a space-ship, would not notice that his clock is ticking more slowly, time is passing more slowly for him, he is aging more slowly and the length of the ruler or measuring device on his space-ship (moving frame) has contracted in the direction of motion (because there is nothing to compare with). According to the Special Theory of Relativity, when this traveler on the space-ship traveling at almost the speed of light (moving frame) looks at a clock on Earth (stationary frame) he would perceive that the clock has slowed down and when he looks at a ruler or measuring device on Earth (stationary frame) he would perceive that it is shorter. But, this is evidently only an illusion and not true, and, the clock ticking away on Earth (stationary frame) is actually ticking more quickly (which implies that time is passing more quickly) as compared to the traveler's clock on the space-ship traveling at almost the speed of light (moving frame) and the ruler or measuring device on Earth (stationary frame) is actually longer as compared to the traveler's ruler or measuring device on the space-ship traveling at almost the speed of light (moving frame) - according to the Special Theory of Relativity, the clock on the space-ship traveling at almost the speed of light (moving frame) has slowed down and both the length of the space-ship traveling at almost the speed of light (moving frame) and the length of the ruler or measuring device on this space-ship (moving frame) have contracted in length in the direction of motion, whilst the clock on Earth (stationary frame) has not slowed down and the ruler or measuring device on Earth (stationary frame) has not contracted in length.

According to the Special Theory of Relativity, the person on Earth (stationary frame) who takes a peep at the clock on the space-ship traveling at almost the speed of light (moving frame) would perceive that the clock ticks more slowly than his clock on Earth (stationary frame). He would also perceive that the ruler or measuring device on the space-ship traveling at almost the speed of light (moving frame) is shorter than his ruler or measuring device on Earth (stationary frame), and, that the space-ship is foreshortened in the direction of its motion. The person on Earth (stationary frame), according to the Special Theory, would see the clock on the space-ship traveling at almost the speed of light (moving frame) slowed down to the same degree as the traveler on the space-ship traveling at almost the speed of light (moving frame) sees the clock on Earth (stationary frame) slowed down and see the length of the ruler or measuring device on the space-ship traveling at almost the speed of light (moving frame), as well as the space-ship's length, shortened to the same degree as the traveler on the space-ship traveling at almost the speed of light (moving frame) sees the ruler or measuring device on Earth (stationary frame) shortened. That is, each of the them, the person on Earth (stationary frame) and the traveler on the space-ship traveling at almost the speed of light (moving frame), according to the Special Theory, would measure the above-mentioned differences in the other to the same degree, and, the other thing each of them would agree on is the constancy of the speed of light (the speed of light would remain constant at 186,000 miles per second).

But, actually, on Earth (stationary frame), the clock has not slowed down and time passes more quickly, and, the length of the ruler or measuring device has not contracted, whereas on the space-ship traveling at almost the speed of light (moving frame) the reverse, as described above, is true. The dilemma is which time is to be regarded as the actual or correct time - the time on Earth (stationary frame) or the time on the space-ship traveling at almost the speed of light (moving frame) - and, which length is to be regarded as the actual or correct length - the length of the ruler or measuring device on Earth (stationary frame) or the length of the ruler or measuring device on the space-ship traveling at almost the speed of light (moving frame)?

According to Einstein, time is relative, i.e., there is no such thing as absolute or actual time, which is a Newtonian concept; Einstein had done away with Newton's concept of absolute time and space. Time depends on the speed of the clock, and, the brain and bodily functions - the nearer the speed of light is approached the slower would be the clock and the passage of time, as well as the brain and bodily functions. What does all this imply? Without a clock, and, consciousness it would not be possible to tell the time - time would not exist, i.e., the reality of time is not independent of the clock, and, consciousness. In fact, without the clock, or, watch everyone would be disoriented where the time is concerned.

However, like the case of the car driver misled by his car's faulty speedometer which is described above, the traveler on the space-ship traveling at almost the speed of light (moving frame) described above has also been misled into thinking that his clock is ticking normally and his time is passing normally, and, his ruler or measuring device is normal and gauging lengths or distances normally. It is only after taking a peep at the clock ticking away and the ruler or measuring device on Earth (stationary frame) that he would realize this might not be so. After having taken a peep at the clock ticking away and the ruler or measuring device on Earth (stationary frame) and perceiving that the clock on Earth (stationary frame) is ticking more slowly and the ruler or measuring device there is shorter (which is in accordance with the Special Theory of Relativity), he should suspect that either his clock on his space-ship traveling at almost the speed of light (moving frame) or the clock on Earth (stationary frame), and, either his ruler or measuring device on his space-ship (moving frame) or the ruler or measuring device on Earth (stationary frame) are not quite right. If he finds out that in fact his clock on his space-ship traveling at almost the speed of light (moving frame) has been ticking more slowly, time has been passing more slowly for him, he has been aging more slowly and the length of the ruler or measuring device on his space-ship traveling at almost the speed of light (moving frame) has contracted in the direction of his space-ship's motion, due to some change in the physical environment, viz., the presence of an intense gravitational field which is created through travel at almost the speed of light, he would realize that the accuracy of the time presented by his clock on his space-ship traveling at almost the speed of light (moving frame), as well as the length or distance measurements made by the ruler or measuring device on his space-ship (moving frame), are out. He would then think that perhaps the time presented by the clock ticking away on Earth (stationary frame) and the length or distance measurements made by the ruler or measuring device there are the actual or correct time and length or distance measurements. But, he might not be able to find out that his clock has been ticking more slowly, time has been passing more slowly for him, he has been aging more slowly and the length of the ruler or measuring device on his space-ship traveling at almost the speed of light (moving frame) has contracted, due to some change in the physical environment, viz., the presence of an intense gravitational field, and, even if he found out that the clock on Earth (stationary frame) is actually ticking more quickly than his and the ruler or measuring device there has actually not contracted in length and is actually longer than his (which is in accordance with the Special Theory of Relativity) he might not understand why and might be puzzled as to whether his clock and ruler or measuring device on his space-ship traveling at almost the speed of light (moving frame) or the clock and ruler or measuring device on Earth (stationary frame) are the clock and ruler or measuring device which are faulty (assuming he has no knowledge of the Special Theory of Relativity).

If neither of the two parties (on Earth (stationary frame) and on the space-ship traveling at almost the speed of light (moving frame)) had been aware that their respective clocks had been ticking away at different speeds, and, the lengths of their respective rulers or measuring devices had been different, then each of them would have regarded the times shown by their respective clocks as the actual time and the lengths displayed by their respective rulers or measuring devices as the actual length, in which case there would be two sets of actual time and actual length, i.e., two sets of reality, a quite absurd situation.

As the third party looking on at the two cases described above and being aware of the circumstances, we could regard the time presented by the clock on Earth (stationary frame) as the actual or correct time and the time presented by the clock on the space-ship traveling at almost the speed of light (moving frame) as the distorted time. Also, as the third party who is all too familiar with the Special Theory of Relativity we could regard the length of the ruler or measuring device on Earth (stationary frame) as the actuality and the length of the ruler or measuring device on the space-ship traveling at almost the speed of light (moving frame), which has contracted in the direction of the space-ship's motion, as the distorted length.

We here consider the example of two space-ships traveling almost next to one another in the same direction, one (we call it X, which is a moving frame) traveling at almost the speed of light, say, 185,000 miles per second (as gauged from Earth, a stationary frame) and the other (we call it Y, which is another moving frame) traveling also at almost the speed of light, say, 185,500 miles per second (as gauged from Earth, a stationary frame). (Theoretically, no space-ship could travel at the speed of light - the Special Theory of Relativity stipulates that at the speed of light everything would be at a standstill - the mass of the space-ship would be infinite and the space-ship would not be able to accelerate anymore, the space-ship's length would have shrunk to zero and any clock within the space-ship would have stopped beating, registering zero time.) The speed of Y (moving frame) as gauged from X (moving frame) or vice versa is computed by using the following formula, as is stipulated by the Special Theory of Relativity:-

$$v = \frac{b - a}{1 - ba/c^2}$$

where c = speed of light = 186,000 miles per second, b = speed of Y = 185,500 miles per second (= 0.9973118c), a = speed of X = 185,000 miles per second (= 0.9946236c)

$$\therefore v = \frac{0.9973118c - 0.9946236c}{1 - 0.9973118c \times 0.9946236c/c^2}$$

$$= \frac{0.0026882c}{1 - 0.9919498c^2/c^2}$$

$$= \frac{0.0026882c}{0.0080502}$$

$= 0.3339295c$ (33.39295 % of speed of light)

$= 0.3339295 \times 186{,}000$ miles per second

$= 62{,}110.887$ miles per second

∴ speed of Y as gauged from X = plus 62,110.887 miles per second (and not plus 500 miles per second (185,500 miles per second minus 185,000 miles per second), which should normally be the case - Y (moving frame) should appear to the traveler in X (moving frame) to be moving away from X (moving frame) in the same direction)

∴ speed of X as gauged from Y = minus 62,110.887 miles per second (and not minus 500 miles per second (minus (185,500 miles per second minus 185,000 miles per second)), which should normally be the case - X (moving frame) should appear to the traveler in Y (moving frame) to be moving away from Y (moving frame) in the opposite direction)

What would be X's and Y's respective speeds then (when gauged from the other), when X (moving frame) and Y (moving frame) travel in opposite directions (instead of the same direction)? The speed of Y (moving frame) as gauged from X (moving frame) and the speed of X (moving frame) as gauged from Y (moving frame) should each not exceed 186,000 miles per second, the speed of light, which represents the ultimate limit, the maximum possible speed any accelerating object could attain, as is stipulated by the Special Theory of Relativity (and not respectively 370,500 miles per second (185,000 miles per second plus 185,500 miles per second), which should normally be the case), and, they are computed by using the following formula (which is described further on), which is in accordance with the Special Theory of Relativity:-

$$v = \frac{a + b}{1 + ab/c^2}$$

where c = speed of light = 186,000 miles per second, a = speed of X = 185,000 miles per second (= 0.9946236c), b = speed of Y = 185,500 miles per second (= 0.9973118c)

$$\therefore v = \frac{0.9946236c + 0.9973118c}{1 + 0.9946236c \times 0.9973118c/c^2}$$

$$= \frac{1.9919354c}{1 + 0.9919498c^2/c^2}$$

$$= \frac{1.9919354c}{1.9919498}$$

$$= 0.9999927c \ (99.99927 \ \% \ of \ speed \ of \ light)$$

$$= 0.9999927 \times 186,000 \ miles \ per \ second$$

$$= 185,998.64 \ miles \ per \ second$$

$\therefore$ speed of Y as gauged from X = speed of X as gauged from Y = 185,998.64 miles per second (Y (moving frame) should appear to the traveler in X (moving frame) to be moving towards X (moving frame) in the opposite direction, and, X (moving frame) should appear to the traveler in Y (moving frame) to be moving towards Y (moving frame) in the opposite direction)

How strange and counter-intuitive it is to find the speeds of X (moving frame) and Y (moving frame) to be minus 62,110.887 miles per second and plus 62,110.887 miles per second respectively as gauged from Y (moving frame) and X (moving frame) respectively (and not minus 500 miles per second and plus 500 miles per second respectively, which should normally be the case) in the first case above, and, to be each only 185,998.64 miles per second (less than the speed of light (186,000 miles per second) and not respectively 370,500 miles per second (185,000 miles per second plus 185,500 miles per second), which should normally be the case) in the second case above, one may think. Evidently, the respective clocks in X (moving frame) and Y (moving frame) were slowing down at different speeds and the respective rulers or measuring devices in X (moving frame) and Y (moving frame) were contracting in length to different extents, since the respective speeds of X (moving frame) and Y (moving frame) are different, viz., 185,000 miles per second and 185,500 miles per second respectively (the higher the speed of the space-ship the more its clock would slow down and the more the length of its ruler or measuring device would contract). As stipulated by the Special Theory of Relativity, both the travelers in X (moving frame) and Y (moving frame) would each see the other's clock as being slower to the same degree and the other's ruler or measuring device as being shorter to the same degree. The dilemma here is to decide whether the clock on X (moving frame) or the clock on Y (moving frame) is giving the correct reading in time and whether the ruler or measuring device on X (moving frame) or the ruler or measuring device on Y (moving frame) is providing the correct measurement in the distance traveled/measured. It is evidently very difficult to decide thus. The travelers in X (moving frame) and Y (moving frame) would each naturally think that everything is fine and consider their respective gauging of the other's speed as correct (assuming that they have no knowledge at all about the Special Theory of Relativity). However, if the travelers in X (moving frame) and Y (moving frame) noticed that the other's clock had been slower and the other's ruler or measuring devices had been shorter, they might each be puzzled and might each wonder whether whose clock and ruler or measuring device are accurate (assuming that they have no knowledge of the Special Theory of Relativity). Of course, if they had known the principles behind the Special Theory of Relativity they would have realized that this phenomenon had been the result of "distortion" due to the creation of an intense gravitational field through travel at almost the speed of light, i.e., the slowing down of their respective clocks and the contraction in the lengths of their respective rulers or measuring devices are transient (they are not permanent - X's and Y's respective clocks would beat at the normal rate and the lengths of their respective rulers or measuring devices would return to their original length once the speeds of X (moving frame) and Y (moving frame) have returned from almost the speed of light (185,000 miles per second and 185,500 miles per second respectively) to the normal speeds, according to the Special Theory of Relativity).

The important question is if the space-ships', X's and Y's, times and length or distance measurements are "distorted" or not real, what should be the real time and real length or distance measurement? However, to both the travelers in space-ships X (moving frame) and Y (moving frame), without being able to compare or look at one another's clock and ruler or measuring device and without any knowledge of the Special Theory of Relativity, the time given by their respective clock and the length or distance measurement given by their respective ruler or measuring device would be the *real* time and *real* length or distance measurement. But, to us, the third party looking on at these two scenarios, who have knowledge of the Special Theory of Relativity, both X's and Y's "real" times and "real" length or distance measurements are illusions and are indeed not real, and, the real time and real length or distance measurement would be those read off a clock and a ruler or measuring device on Earth (stationary frame), where the clock and the ruler or measuring device are free from the "distortional" effect of the intense gravitational field created through travel at almost the speed of light. Since to the traveler on space-ship X (moving frame), the traveler on space-ship Y (moving frame) and to us on Earth (stationary frame), our times and length or distance measurements are "real" to each of us, it is implied that time and length or distance measurement are relative, i.e., time and length or distance measurement depend on environmental or situational factors.

All this appears to be a case of how we choose to interpret these three scenarios. For example, if we put ourselves in the X traveler's shoes, have no knowledge that Y (moving frame) exists or if we know that Y (moving frame) exists we have no knowledge that Y's clock is running at a different pace and that Y's ruler or measuring device is gauging length or distance differently from our ruler or measuring device, and, have no knowledge that our clock has slowed down and that our ruler or measuring device has shrunk in length, i.e., we have no knowledge of the Special Theory of Relativity, then we would just think that our time and length or distance measurement in X (moving frame) are the real things. The same applies if we put ourselves in the Y traveler's shoes. On the other hand, if the person concerned were wised up to the Special Theory of Relativity he could choose to regard the real time and real length or distance measurement as those only made by any clock and any ruler or measuring device on Mother Earth (stationary frame). This is only one possible interpretation. A philosophical-minded person could choose the other interpretation, viz., X's time and length or distance measurement are real to the traveler in X (moving frame), Y's time and length or distance measurement are real to the traveler in Y (moving frame), and, Earth's time and length or distance measurement are real to the resident on Earth (stationary frame), i.e., there are different realities. Should there be only one reality or should more than one reality be allowed?

Therefore, consciousness or knowledge of certain facts, e.g., the fact that there is another clock under a different set of circumstances ticking away at a different speed, as described above, or, simply some other clock to compare the time with would affect our sense of time, time being the fourth dimension in Einstein's Relativity theory.

Also, a person could be deceived about the time by a faulty clock, a case which is similar to the above-mentioned case of the car driver being misled by the faulty speedometer of his car.

All this implies that time is subjective, or, as Einstein had put it, relative (not absolute), depending on the situation, and, consciousness has an important role to play.

There should be a sufficient reason to explain why the clock, and, the brain and bodily functions of the person slow down, and the length of the ruler or measuring device contracts, on approaching the speed of light, while at the speed of light the mechanism of the clock and time are at a standstill and the length of the ruler or measuring device is zero, which is important. Though the intense gravitational field caused by travel at almost the speed of light might account for the slowing down of the clock (for which experimental evidence had been obtained) and therefore time, as well as the brain and bodily functions of a person, it evidently hardly suffices as an explanation for the contraction of the length of the ruler or measuring device in the direction of travel at almost the speed of light (for which experimental evidence has yet to be found, and, which seems like a "fudge on the figure" by the inventor of the theory to "ensure the constancy of the speed of light"). Though the constancy of the speed of light as gauged from the Earth is evidently a well-proven phenomenon, no one has yet been able to travel at almost the speed of light and gauge the speed of a light beam by traveling besides it in the same direction, as described above - despite the experimental findings that at high speeds, though very much less than the speed of light, clocks slow down, the contraction of rulers or measuring devices in the direction of motion at almost the speed of light is evidently only an inference, with no experimental basis.

The following equation describes how the speed of light (v) is derived:-

$$v = d/t,$$

where d represents the distance traveled by the light beam and t represents the time taken by the light beam to travel the distance d

Since time is relative (and not absolute) and depends on the mechanism of the clock, as well as consciousness, which slow down on approaching the speed of light, it could be arbitrary. The clock which is used to gauge the time t taken by the light beam to travel the distance d might not slow down uniformly (at the same rate) on approaching the speed of light (under normal, earthly conditions time varies from clock to clock by minutes or more - there is evidently some uncertainty in the mechanism of clocks). Besides, the ruler or measuring device used to gauge the distance d traveled by the light beam in time t might not contract in length uniformly (at the same rate) on approaching the speed of light. If the clock does not slow down uniformly (at the same rate) and the ruler or measuring device does not contract in length uniformly (at the same rate) on approaching the speed of light there is all probability that the speed of light (v) as represented by d/t would be variable, higher than 186,000 miles per second at times, below 186,000 miles per second at other times, or, equal to 186,000 miles per second at yet other times. Moreover, in accordance with the Special Theory of Relativity, as described above, for the speed of light to really remain constant, on approaching the speed of light the clock must slow down to the same degree as the contraction in the length of the ruler or measuring device. (This explanation presented by the Special Theory of Relativity is incorrect.) We describe these possible outcomes as follows. (Only for argument's sake here, we assume that the above statement "for the speed of light to really remain constant, on approaching the speed of light the clock must slow down to the same degree as the contraction in the length of the ruler or measuring device" is correct and would lead to the following possible outcomes; the examples which follow, for the sake of argument, would also be based on this assumption.):-

$$\text{i) } S^\% > C^\% \rightarrow I^l$$
$$\text{ii) } S^\% < C^\% \rightarrow D^l$$
$$\text{iii) } S^\% = C^\% \rightarrow S^l$$

where $S^\%$ represents percentage of slowing down of the clock, $C^\%$ represents percentage of contraction in the length of the ruler or measuring device, I^l represents increase in the speed of light, i.e., exceed 186,000 miles per second, D^l represents decrease in the speed of light, i.e., go below 186,000 miles per second, S^l represents speed of light, i.e., 186,000 miles per second

We ponder this point more deeply by reconsidering the two examples pertaining to space-ships X (moving frame) and Y (moving frame) which have been described above.

Let us look at the first case pertaining to space-ships X (moving frame) and Y (moving frame) traveling at speeds of 185,000 miles per second (as gauged from Earth, a stationary frame) and 185,500 miles per second (as gauged from Earth, a stationary frame) respectively in the same direction almost next to one another. The speed of Y (moving frame) as gauged from X (moving frame) should normally be plus 500 miles per second (185,500 miles per second minus 185,000 miles per second) and the speed of X (moving frame) as gauged from Y (moving frame) should normally be minus 500 miles per second (minus (185,500 miles per second minus 185,000 miles per second)), as explained above. But, the speed of Y (moving frame) as gauged from X (moving frame) and the speed of X (moving frame) as gauged from

Y (moving frame) should be plus 62,110.887 miles per second and minus 62,110.887 miles per second respectively, as computed by using the formula below, which is in accordance with the Special Theory of Relativity:-

$$v = \frac{b - a}{1 - ba/c^2}$$

where c = speed of light = 186,000 miles per second, b = speed of Y = 185,500 miles per second (= 0.9973118c), a = speed of X = 185,000 miles per second (= 0.9946236c)

$$= 62,110.887 \text{ miles per second}$$

speed of Y as gauged from X = plus 62,110.887 miles per second (and not plus 500 miles per second (185,500 miles per second minus 185,000 miles per second), which should normally be the case - Y (moving frame) should appear to the traveler in X (moving frame) to be moving away from X (moving frame) in the same direction)

speed of X as gauged from Y = minus 62,110.887 miles per second (and not minus 500 miles per second (minus (185,500 miles per second minus 185,000 miles per second)), which should normally be the case - X (moving frame) should appear to the traveler in Y (moving frame) to be moving away from Y (moving frame) in the opposite direction)

Evidently, the clocks and rulers or measuring devices on X (moving frame) and Y (moving frame) have respectively slowed down and contracted in length in the direction of motion to different degrees while X (moving frame) and Y (moving frame) have been traveling at 185,000 miles per second and 185,500 miles per second respectively, almost the speed of light. Whilst the respective speeds of Y (moving frame) and X (moving frame) as gauged from the other should be plus 500 miles per second and minus 500 miles per second respectively, the slowing down of their respective clocks and the contraction in the lengths of their respective rulers or measuring devices to different degrees have resulted in the above-mentioned plus/minus 500 miles per second being gauged differently, as shown in the following examples:-

For example, if the length of the ruler or measuring device on the space-ship contracts by 20 % while the space-ship travels at almost the speed of light, the relative speed of plus/minus 500 miles per second of each of the space-ships, X (moving frame) and Y (moving frame), should be recomputed/gauged as follows to produce the "distorted" speed of plus/minus

62,110.887 miles per second, which is in accordance with the Special Theory of Relativity:

$$\frac{62,110.887 \text{ miles}}{1 \text{ second}} =$$ 400 miles (0.8 of 500 miles - due to ruler length contraction of **20 %**, 500 miles are gauged by space-ship traveler as 400 miles) ÷ 0.00644 second (due to clock on space-ship slowing down by **15528 %**, 1 second is gauged by space-ship traveler as 0.00644 second)

For example, if the length of the ruler or measuring device on the space-ship contracts by 40 % while the space-ship travels at almost the speed of light, the relative speed of plus/minus 500 miles per second of each of the space-ships, X (moving frame) and Y (moving frame), should be recomputed/gauged as follows to produce the "distorted" speed of plus/minus 62,110.887 miles per second, which is in accordance with the Special Theory of Relativity:

$$\frac{62,110.887 \text{ miles}}{1 \text{ second}} =$$ 300 miles (0.6 of 500 miles - due to ruler length contraction of **40 %**, 500 miles are gauged by space-ship traveler as 300 miles) ÷ 0.00483 second (due to clock on space-ship slowing down by **20703.93 %**, 1 second is gauged by space-ship traveler as 0.00483 second)

For example, if the length of the ruler or measuring device on the space-ship contracts by 60 % while the space-ship travels at almost the speed of light, the relative speed of plus/minus 500 miles per second of each of the space-ships, X (moving frame) and Y (moving frame), should be recomputed/gauged as follows to produce the "distorted" speed of plus/minus 62,110.887 miles per second, which is in accordance with the Special Theory of Relativity:

$$\frac{62,110.887 \text{ miles}}{1 \text{ second}} =$$ 200 miles (0.4 of 500 miles - due to ruler length contraction of **60 %**, 500 miles are gauged by space-ship traveler as 200 miles) ÷ 0.00322 second (due to clock on space-ship slowing down by **31055.9 %**, 1 second is gauged by space-ship traveler as 0.00322 second)

However, if the length of the ruler or measuring device on the space-ship contracts by, e.g., 20 %, and the clock on the space-ship slows down by the same percentage, i.e., 20 %, while the space-ship travels at almost the speed of light, the relative speed of plus/minus 500 miles per second of each of the space-ships, X (moving frame) and Y (moving frame), would remain the same, i.e., plus/minus 500 miles per second, which is in accordance with the Special Theory of Relativity:

$\dfrac{500 \text{ miles}}{1 \text{ second}}$ = 400 miles (0.8 of 500 miles - due to ruler length contraction of **20 %, 5**00 miles are gauged by space-ship traveler as 400 miles) ÷ 0.8 second (due to clock on space-ship slowing down by **20 %**, 1 second is gauged by space-ship traveler as 0.8 second)

We now look at the second case pertaining to space-ships X (moving frame) and Y (moving frame) traveling at speeds of 185,000 miles per second (as gauged from Earth, a stationary frame) and 185,500 miles per second (as gauged from Earth, a stationary frame) respectively in opposite directions. The speed of Y (moving frame) as gauged from X (moving frame) and the speed of X (moving frame) as gauged from Y (moving frame) should each normally be 370,500 miles per second (185,000 miles per second plus 185,500 miles per second), as explained above. However, the speed of Y (moving frame) as gauged from X (moving frame) and the speed of X (moving frame) as gauged from Y (moving frame) should each not exceed 186,000 miles per second, the speed of light, which represents the ultimate limit, the maximum possible speed any accelerating object could attain, as stipulated by the Special Theory of Relativity (and not respectively 370,500 miles per second (185,000 miles per second plus 185,500 miles per second)), which should normally be the case), and, they are computed by using the following formula, which is in accordance with the Special Theory of Relativity:-

$$v = \frac{a + b}{1 + ab/c^2}$$

where c = speed of light = 186,000 miles per second, a = speed of X = 185,000 miles per second (= 0.9946236c), b = speed of Y = 185,500 miles per second (= 0.9973118c)

$$= 185,998.64 \text{ miles per second}$$

speed of Y as gauged from X = speed of X as gauged from Y = 185,998.64 miles per second (Y (moving frame) should appear to the traveler in X (moving frame) to be moving towards X (moving frame) in the opposite direction, and, X (moving frame) should appear to the traveler in Y (moving frame) to be moving towards Y (moving frame) in the opposite direction)

Let us look at the following examples:-

For example, if the length of the ruler or measuring device on the space-ship contracts by 60 % while the space-ship travels at almost the speed of light, the relative speed of 370,500 miles per second of each of the space-ships, X (moving frame) and Y (moving frame), should be recomputed/gauged as follows to produce the "distorted" speed of 185,998.64 miles per second, which is in accordance with the Special Theory of Relativity:

$$\frac{185,998.64 \text{ miles}}{1 \text{ second}} = 148,200 \text{ miles } (0.4 \text{ of } 370,500 \text{ miles - due to ruler}$$

length contraction of **60 %**, 370,500 miles are gauged by space-ship traveler as 148,200 miles) ÷ 0.79678 second (due to clock on space-ship slowing down by **125.51 %**, 1 second is gauged by space-ship traveler as 0.79678 second)

For example, if the length of the ruler or measuring device on the space-ship contracts by 70 % while the space-ship travels at almost the speed of light, the relative speed of 370,500 miles per second of each of the space-ships, X (moving frame) and Y (moving frame), should be recomputed/gauged as follows to produce the "distorted" speed of 185,998.64 miles per second, which is in accordance with the Special Theory of Relativity:

$$\frac{185,998.64 \text{ miles}}{1 \text{ second}} = 111,150 \text{ miles } (0.3 \text{ of } 370,500 \text{ miles - due to ruler}$$

length contraction of **70 %**, 370,500 miles are gauged by space-ship traveler as 111,150 miles) ÷ 0.597585 second (due to clock on space-ship slowing down by **167.34 %**, 1 second is gauged by space-ship traveler as 0.597585 second)

For example, if the length of the ruler or measuring device on the space-ship contracts by 80 % while the space-ship travels at almost the speed of light, the relative speed of 370,500 miles per second of each of the space-ships, X (moving frame) and Y (moving frame), should be recomputed/gauged as follows to produce the "distorted" speed of 185,998.64 miles per second, which is in accordance with the Special Theory of Relativity:

$\dfrac{185,998.64 \text{ miles}}{1 \text{ second}}$ = 74,100 miles (0.2 of 370,500 miles - due to ruler length contraction of **80 %**, 370,500 miles are gauged by space-ship traveler as 74,100 miles) ÷ 0.39839 second (due to clock on space-ship slowing down by **251.01 %**, 1 second is gauged by space-ship traveler as 0.39839 second)

However, if the length of the ruler or measuring device on the space-ship contracts by, e.g., 60 %, and the clock on the space-ship slows down by the same percentage, i.e., 60 %, while the space-ship travels at almost the speed of light, the relative speed of 370,500 miles per second of each of the space-ships, X (moving frame) and Y (moving frame), would remain the same, i.e., 370,500 miles per second, which is in accordance with the Special Theory of Relativity:

$\dfrac{370,500 \text{ miles}}{1 \text{ second}}$ = 148,200 miles (0.4 of 370,500 miles - due to ruler length contraction of **60 %**, 370,500 miles are gauged by space-ship traveler as 148,200 miles) ÷ 0.4 second (due to clock on space-ship slowing down by **60 %**, 1 second is gauged by space-ship traveler as 0.4 second)

Thus, as is evident from the above examples, which are in accordance with the Special Theory of Relativity, to arrive at the two speeds, i.e., 62,110.887 miles per second and 185,998.64 miles per second, as well as other speeds, obtained by using the formulas stipulated by the Special Theory of Relativity, $v = (b - a) \div (1 - ba/c^2)$ and $v = (a + b) \div (1 + ab/c^2)$, the clocks and the rulers or measuring devices on the space-ships traveling at almost the speed of light would have to each respectively slow down and contract in length **at different rates** (and definitely not at the same rate). The only exception is evidently the case of the constancy of the speed of light, whereby the clock and the ruler or measuring device have to each respectively slow down and contract in length **at the same rate**, giving the **same percentage** decrease in the time gauged and the distance gauged, as follows, as stipulated by the Special Theory of Relativity:

(186,000 miles - X % of 186,000 miles) ÷ (1 second - X % of 1 second) = 186,000 miles per second

Why is the constancy of the speed of light the **exception**? Was it an adjustment or modification of the mathematics to "ensure" the constancy of the speed of light? Could the speed of light not be variable, below, at and above 186,000 miles per second at various times, as some have suggested?

Let us look again at the case of the constancy of the speed of light. We have stated that according to the Special Theory of Relativity the speed of a beam of light (c = 186,000 miles per second) would always be found to be constant or unchanged when gauged by a person traveling close to the beam of light in the same direction at almost the same speed as the beam of light (say 0.9c = 167,400 miles per second) because the clock and ruler or measuring device used by the person traveling in the same direction at almost the speed of light in gauging the speed of the beam of light would have respectively slowed down and contracted in length **at the same rate** (say X %).

We have to remember that the speed of the beam of light is obtained by dividing the distance traveled by the beam of light (as gauged by the ruler or measuring device on the space-ship traveling at almost the speed of light - moving frame, which has contracted in length) by the time it took to travel that distance by the beam of light (as gauged by the clock therein the space-ship - moving frame, which has slowed down), the gauging being carried out by the traveler on the space-ship (moving frame). Let us, using a simple example, say that the one-metre-long ruler used to gauge distance has contracted in length in the direction of motion by 20 %. The clock used to gauge time, which has also slowed down by 20 %, according to the Special Theory of Relativity, would now gauge the time taken, say X, to travel the distance between two designated points (reference, stationary frame), say Y, as having decreased by 20 % to become 0.8 X. Though the one-metre-long ruler, which has contracted in length by 20 %, still reads "1 metre" in length, it is in effect shorter by 20 % (actually only 0.8 metre in length). Therefore, when it gauges the distance traveled, Y, above, this distance Y would now be gauged as 1.25 Y, and not 0.8 Y in accordance with the Special Theory of Relativity. As stated above, Special Relativity theorizes that for the speed of a beam of light to remain constant the beam of light would have to take less time (time dilation) to travel a shorter distance (length contraction) - in effect, X % less time to travel a distance shorter by X %, in accordance with the above-mentioned equation, which implies that the speed of the beam of light would remain constant, e.g., 0.8 X (time) to travel 0.8 Y (distance) after "time dilation" and "length contraction". But, as explained above, this would not be the case; the beam of light would have been gauged as having taken 0.8 X (time) to travel 1.25 Y (distance). This is an anomaly and it shows that there is something not right with the Special Theory of Relativity.

Since light particles (photons) do not have mass or inertia, which prevents an object possessing it from accelerating beyond the speed of light, viz., 186,000 miles per second, theoretically there is nothing to prevent light particles (photons) or other objects without mass or inertia from traveling at a speed greater than 186,000 miles per second.

What would be the speed of the beams of light from the head-lights of a vehicle traveling at, e.g., 0.0083 mile per second, or, 30 miles per hour? If gauged from outside the moving vehicle (stationary frame) it should normally be 186,000.0083 miles per second (speed of the moving vehicle (0.0083 mile per second, or, 30 miles per hour) plus speed of each beam of light (186,000 miles per second)), but, according to the Special Theory of Relativity, this is not the case and the speed of the beams of light is not 186,000.0083 miles per second but still 186,000 miles per second (constant). How come? This is because light travels on its own independently of or unaffected by its source, in the above case, the source of the beams of light being the head-lights of the vehicle traveling at 0.0083 mile per second, or, 30 miles per hour.

In the above case, the beams of light are so much faster (186,000 miles per second) than the speed of the moving vehicle (0.0083 mile per second, or, 30 miles per hour) that they would be continually moving way way ahead of the vehicle after they leave the head-lights. Logically, if the vehicle had traveled at a speed exceeding the speed of light (which according to the Special Theory of Relativity is impossible - no object could travel faster than the speed of light - it is only an assumption here for the sake of argument) and continually "overtaken" the beams of light that its head-lights had emitted, the beams of light (whose speed remains the same at 186,000 miles per second - constant) should not be in front of the head-lights. If the vehicle were traveling much in excess of the speed of light, the beams of light emitted from the vehicle's head-lights should tag behind the vehicle. This is counter to our normal experiences with light beams from the head-lights of moving vehicles, which always appear in front of the vehicles.

The following equation shows that no moving object could travel faster than the speed of light, which is in accordance with the Special Theory of Relativity:-

$$v = \frac{a + b}{1 + ab/c^2}$$

If we let a = velocity of moving train, b = velocity of light beam (which is sent from the back of the moving train to the front of the moving train) with respect to the moving train, which is the moving frame (i.e., the velocity of the light beam (which is sent from the back of the moving train to the front of the moving train) is gauged from the moving train, which is the moving frame), v = velocity of light beam (which is sent from the back of the moving train to the front of the moving train) with respect to the ground level, which is the stationary frame (i.e., the velocity of the light beam (which is sent from the back of the moving train to the front of the moving train) is gauged from the ground level, which is the stationary frame), c = velocity of light = 186,000 miles per second, and, also let a = b = c, then:-

$$v = \frac{c + c}{1 + c.c/c^2} = 2c/2 = c! \text{ (And not 2c!)}$$

Though theoretically no object could travel faster than the speed of light because at the speed of light the object's mass is infinitely great and therefore it is unable to accelerate, an object without mass, possibly, a quantum particle which is somewhat similar to a photon (a photon is a quantum particle without mass always in motion) might be capable of traveling faster than the speed of light. Such an object or objects might be waiting to be discovered. As it is, a "theoretical" particle which travels faster than the speed of light, which is termed "tachyon", has been thought to exist.

There have been a number of speculations pertaining to the variable speed of light (VSL), e.g., one theory states that the speed of light varies with the various stages of the evolution of the universe, exceeding 186,000 miles per second at certain points of time.

In the Special Theory of Relativity it is implied that an object traveling at more than the speed of light would go backwards in time, which is bizarre, but has evidently been taken seriously by quite a number. If time-travel were indeed possible, a person could go back to the time before he was born and murder his grandfather so that his father, and, hence, himself, would not have been conceived, which is against causality, paradoxical and absurd, thus implying that time-travel is not really possible.

After an object had contracted to zero length at the speed of light, it would actually cease to exist. Could an object really exceed the speed of light, contract in length further to "minus something" length from zero length or non-existence and travel backwards in time, as have been postulated by a number of people who hold the view that time-travel is possible? Negative quantities, e.g., negative lengths, as such, are evidently abstractions devoid of any real meaning or existence. For instance, a string which is minus one foot in length or a boat which is minus ten feet in length is meaningless. The idea of time-travel suggested by the Special Theory of Relativity should be only regarded as a metaphor or curiosity; it should be regarded as something which might be theoretically possible but is not practicable. It is somehow comparable to one of those well-known paradoxes, e.g., Zeno's paradox whereby it is shown that Achilles would never be able to overtake a tortoise which was given a head-start in a race (which is absurd).

Stephen Hawking has stated that if time-travel were possible we would have had visitations from tourists from the future (which we have evidently none so far), which, according to him, shows that time-travel is not possible. But, what about the reports and books about alleged sightings of UFOs and encounters with or abductions by aliens or extra-terrestrials? Are these alleged aliens or extra-terrestrials not possible tourists from the future, beings with seemingly advanced technologies, technologies which seem more advanced than ours, e.g., flying saucers, instantaneous appearance and disappearance which are suggestive of teleportation, et al.? It is also possible, and seemingly much likely, that such beings, if they indeed exist, are from a civilization or civilizations from another part of the universe or another universe, which are more advanced than ours. There have been much stories and speculations pertaining to time-travel or teleportation. It has been discovered that quantum particles could be teleported, i.e., made to appear instantaneously at another location. If only there is a technology to teleport all the atoms in the body of a human being so that the human being could appear instantaneously at another place without physical and mental harm.

The stipulation of the impossibility of exceeding the speed of light by the Special Theory of Relativity means that time-travel is not possible, and Einstein had stated that time-travel is impossible. Though time-travel in the physical dimension is evidently impossible, time-travel in the domain of the consciousness or mind is possible.

3 SPEED OF LIGHT AS THE LIMIT

One of the postulates of the Special Theory of Relativity is that no object could travel at a greater velocity than the velocity of light. However, it should be noted that quantum particles are capable of teleportation or travel to another location instantaneously, which is in effect superluminal or faster than light travel.

The postulates of the Special Theory of Relativity evidently imply that the invariance of the velocity of light at all inertial frames is only an illusion - if the velocity of light were to appear invariant, according to the Theory, lengths have to contract (Lorentz contraction) and clocks have to slow down (time dilation), at the same rate, while traveling at close to the velocity of light. We here ask the important question: If lengths do not contract and clocks do not slow down at close to the velocity of light, as are postulated by the Special Theory of Relativity, would the velocity of light still appear invariant?

In all this, we should also not forget that while the slowing down of clocks (time dilation) when traveling at high velocities is an experimentally proven phenomenon length contraction (Lorentz contraction) has not been experimentally proven and remains an inference - length contraction (Lorentz contraction) might not be an actuality.

In Chapter 1, it is stated that there is an anomaly relating to length contraction (Lorentz contraction) and the invariance of the velocity of light. Thus, there could be other reason or reasons for the invariance of the velocity of light at all inertial frames, e.g., length expansion, as is described in this earlier Chapter (as per Item (1) below), or, some other valid reasons. Let us here recapitulate the important point brought up in this earlier Chapter, which is as follows:-

In order for the velocity of the beam of light to remain/appear invariant, one of the following has to happen:

1) When the clock slows down (time dilation) by x %, the ruler should increase in length (length expansion) by y %.
2) When the ruler decreases in length (length contraction) by x %, the clock should quicken (time contraction) by y %.
3) When the clock slows down (time dilation) by q % and the ruler decreases in length (length contraction) by q %, the beam of light (moving frame) should slow down by r %.

We would add a fourth option to the above three options, which is as follows:

4) When the clock slows down (time dilation) by q % and the ruler remains
   the same in length (unchanged in length) if Lorentz contraction were not an
   actuality and does not happen, the beam of light (moving frame) should
   slow down by s %. (s % < r %)

As is stated above, Special Relativity postulates that no moving object could exceed the velocity of light, as at the velocity of light the moving object's mass would be infinite, making further acceleration or increase in velocity of the moving object impossible, while its length would have shrunk to zero - hence the invariance of the velocity of light. Mass is a primitive concept in mechanics. It is assumed to be additive for disjoint bodies. In Newtonian dynamics, it is constant for a given set of particles that may either constitute a body or be discrete. It is normally measured in kilograms. In practice the mass of a body is found by measuring its weight. Formally, mass is a measure; if the set of particles is a body it is required that the measure of the body be absolutely continuous with respect to Lebesgue measure. If length contraction (Lorentz contraction) of a moving object were indeed a true phenomenon (proven by experiment), then at zero length when moving at the velocity of light, as is postulated by the Special Theory of Relativity, the object would have practically disappeared; this is evidently the second reason why no object could exceed the velocity of light - at the velocity of light the object would practically disappear, being zero in length. The object's mass should have also disappeared at the same time, instead of being infinite, as the Special Theory of Relativity has postulated, as mass is linked to size such as length - no body, no size, no mass, and zero length should mean zero mass. The concept of mass is sticky as it is never measured as size such as length, breadth, height, area or volume and only as weight. Matter has mass and it also has size such as length, breadth, height, area or volume. Size - length, breadth, height, area or volume - is related to mass, normally the larger the size, e.g., length, is the larger the mass would be and vice-versa - heavier objects are normally larger in size and vice-versa. Therefore, length contraction (Lorentz contraction) and the increase in mass as per the postulation of the Special Theory of Relativity do not appear to fit together logically. It appears to be an anomaly. So far no one has been able to explain length contraction (Lorentz contraction) though Lorentz himself did try to explain the phenomenon, which is heretofore merely an inference.

On the other hand, length expansion, as in Item (1) above, goes logically well with the increase in mass as the above explanation shows, i.e., the idea that as an object travels at greater and greater velocity approaching the velocity of light its length expands simultaneously together with the increase in its mass makes better sense. But could Item (1) be seriously considered as the solution to the anomaly pertaining to length contraction (Lorentz contraction) and the invariance of the velocity of light as is described in Chapter 1? Could an object traveling at high velocities possibly increase in length (experience length expansion) and what could cause this to happen?

Since the phenomenon of clocks slowing down (time dilation) while traveling at high velocities had been confirmed by experiments and length contraction has not been confirmed experimentally as yet but is an inference only, Item (2) above (which states clock quickening, an unproven phenomenon and the reverse and contradiction of the experimentally proven clock slowing down phenomenon (time dilation)) could be ruled out, while Items (1), (3) and (4) are possibilities, however remote these possibilities might be. Though the intense gravitational field caused by travel at almost the velocity of light might account for the slowing down of clocks (for which experimental evidence had already been obtained as is stated above) and therefore time, as well as the brain and bodily functions of a person, it evidently hardly suffices as an explanation for length contraction (for which experimental evidence has yet to be found, and, which seems like a "fudge on the figure" by the inventor of the Special Theory of Relativity to "ensure the invariance of the velocity of light").

We should remember that length contraction is after all an unconfirmed inference (unlike time dilation which had been proven by experiments as is stated above). The same would apply to length expansion. There is probably no such things as length contraction or length expansion. It is difficult to envision or imagine a rigid object such as a ruler or metre rod contracting in length or expanding in length as though it is made of rubber, which is flexible, and such a phenomenon should be regarded as improbable; length contraction and length expansion could therefore be regarded as only illusions at most, more apparent than real, like some of the other postulates of the Special Theory of Relativity, which would be described below. Because of this, Items (1) and (3) above would appear remotely probable with Item (2) completely ruled out as is stated above, while Item (4) is most probable. But Item (4) above implies that the velocity of light would appear to exceed the 186,000 miles per second limit (the slowed down clock ("time dilated" clock) and the ruler which remains the same in length (does not contract in length) would now together gauge the velocity of the beam of light (which is actually 186,000 miles per second) as more than 186,000 miles per second - as the clock has slowed down (time dilation), the beam of light would now (appear to) take less time to travel the same distance, i.e., the velocity of the beam of light now appears to be greater, this higher velocity being determined by dividing the distance traveled by the time taken to travel this distance), 186,000 miles per second being the limit of the velocity of light which is postulated by the Special Theory of Relativity - the velocity of light could never exceed this limit as is postulated by the Theory. Thus, the above is evidently an *illusion* caused by the slowing down of the clock while the length of the ruler remains unchanged (does not contract), both the clock and the ruler having been utilised to gauge the velocity of the beam of light. That is, Item (4) above would produce the *illusion* of the beam of light (which has an actual velocity of 186,000 miles per second) having a velocity of more than 186,000 miles per second. All this would be another "headache" for the Special Theory of Relativity, which states that no moving object including light could exceed the velocity limit of 186,000 miles per second. As is stated in Item (4) above, in order for the slowed down clock (slowed down by q % for example) and the ruler whose length has not contracted but remains the same to gauge the velocity of the beam of light as invariant (invariant at 186,000 miles per second), the actual velocity of the beam of light has to be less than 186,000 miles per second (the beam of light should slow down by s %, as is stated in Item (4) above); this would of course result in the *illusion* that the velocity of the beam of light is invariant (unchanged at 186,000 miles per second) while the actual velocity of the beam of light is less than 186,000 miles per second - if the clock used to gauge the velocity of the beam of light (which is actually less than 186,000 miles per second, say d miles per second) had not slowed down (time dilation) but remained ticking at the same rate, the velocity measured would certainly be less than 186,000 miles per second (which is as stated just above the actual velocity, i.e., d miles per second).

However, of the four options above, Items (1), (2), (3) and (4), Item (4) is hence evidently the most realistic and probable. We recapitulate here: Item (4) states that there is time dilation but no length contraction, i.e., clocks would slow down at high velocities, e.g., velocities close to the velocity of light, but at such high velocities rulers would not contract in length in the direction of motion and would remain the same in length. Based on these conditions of Item (4), there is a logical, more sensible explanation for the invariance of the velocity of light, which would be described shortly.

First, the fact that the velocity of light would be invariant if gauged from different inertial frames, e.g., when the velocity is gauged when the beam of light is emitted from a source which is stationary (not moving), for instance, the headlight of a stationary car, and, when the velocity is gauged when the beam of light is emitted from a source which is moving, for instance, the headlight of a moving car - in both these instances the velocity of the beam of light would be the same, as is postulated by the Special Theory of Relativity, though common sense dictates that in the second instance, the instance of the moving car, the velocity of the beam of light should be the velocity of the beam of light (186,000 miles per second) plus the velocity of the car (say 0.014 miles per second), giving a total velocity of 186,000.014 miles per second. The answer to this abnormality, according to the Special Theory of Relativity, is that the beam of light is independent of its source, the car headlight, and is not affected by this source. This implies that if the car were to travel at a higher velocity than the velocity of the beam of light the car would be moving in front of the beam of light, while the beam of light would be tagging behind the car.

Second, a person on a moving vehicle, e.g., a very fast moving train (moving frame), traveling at close to the velocity of a beam of light (moving frame) in the same direction would find the velocity of the beam of light (moving frame) to be invariant at 186,000 miles per second, instead of the difference between the velocity of the very fast moving train (moving frame) and the velocity of the beam of light (moving frame), which would normally be the case. This is because, according to the Special Theory of Relativity, on the very fast moving train (moving frame) approaching the velocity of light the clock therein used to gauge the time traveled by the beam of light (moving frame) has slowed down by the same degree (say X

%) as the ruler or measuring device (stated as meter stick or measuring rod in some texts) therein used to gauge the distance traveled by the beam of light (moving frame) has contracted in length in the direction of the very fast moving train's motion (also X %), the greater the very fast moving train's traveling velocity the more the clock slows down and the greater the length contraction of the ruler or measuring device. This is expressed in the following equation (the velocity of the beam of light (moving frame) being the distance it traveled divided by the time it took to travel this distance), which is in accordance with the Special Theory of Relativity:-

(186,000 miles - X % of 186,000 miles) ÷ (1 second - X % of 1 second) = 186,000 miles per second

The explanation in this second case for the invariance of the velocity of light is based on the conditions stipulated in Item (3) above, wherein length contraction is evidently not so probable, as is explained above. Also, to have length contraction and time dilation happen to the same degree (shown in the above equation as X % each) is not that probable. We would re-construe this second case using the more realistic and probable conditions stated in Item (4) above, namely, time dilation, which is a proven phenomenon, and absence of length contraction, which is to be expected. We would use a simple diagram to explain, which is as follows:-

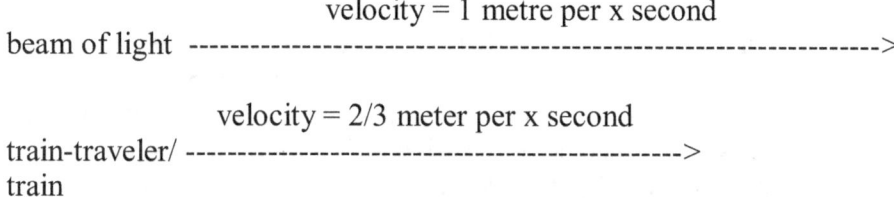

The above train-traveler traveling at two-third the velocity of light besides a beam of light in the same direction is gauging the velocity of the beam of light. The velocity of the beam of light is 1 metre per x second while the velocity of the train, being two-third the velocity of light, is 2/3 metre per x second. Under normal circumstances, the train-traveler would gauge the velocity of the beam of light traveling in the same direction besides his train as 1/3 metre per x second, obtained by deducting the velocity of the train from the velocity of the beam of light, i.e., 1 metre per x second minus 2/3 metre per x second. But, the train-traveler is now experiencing some abnormal conditions while traveling at two-third the velocity of light, i.e., 2/3 metre per x second, as, by Item (4) above, his clock slows down while his ruler or measuring rod remains unchanged in length (does not experience length contraction). Say, e.g., the train-traveler's clock has slowed down by two-third while his ruler or measuring rod remains the same in length, while traveling at two-third the velocity of light, or, 2/3 metre per x second. The train-traveler's unchanged ruler or measuring rod would now gauge 1/3 metre (the velocity of the beam of light should be 1/3 metre per x second under normal circumstances as is stated above) as 1/3 metre still but his clock which has slowed down by two-third would now gauge the time taken to travel the distance of 1/3 metre as 1/3 x second (and not x second), i.e., the train-traveler would now gauge the velocity of the beam of light traveling in the same direction besides his train as 1/3 metre per 1/3 x second, which is the same as 1 metre per x second, which is the velocity of light! Thus, to the train-traveler, the velocity of the beam of light traveling in the same direction besides his train is invariant, i.e., still 1 metre per x second, instead of 1/3 metre per x second. Therefore, the conditions of Item (4) above, namely time dilation and absence of length contraction, could be incorporated into a revised Special Theory of Relativity, whereby the anomalies would be gone.

We have to find a fool-proof reason or reasons to explain why the velocity of light always appears invariant at all inertial frames. If not, the invariance of the velocity of light would be subject to doubt. In the light of the above, we should query whether the invariance of the velocity of light postulated by the Special Theory of Relativity is *real* or just an *illusion*. In fact, there are some well-respected scientists who think that the velocity of light is variable, with one well-respected UK based scientist entertaining the idea that the velocity of light has varied with the evolution of the universe, exceeding 186,000 miles per second at certain points of time.

On the one hand, there had been quite some evidence which prove that the velocity of light is invariant, e.g., the famous Michelson-Morley experiments. On the other hand, this important tenet of Special Relativity has its contradictions or weaknesses. It is a confusing state of affairs and we find many admirers as well as detractors of Special Relativity. In view of all the above-described, including the postulates of the Special Theory of Relativity, the velocity of a beam of light could be regarded as "relative" and "variable", depending on how the measuring devices such as the clock and the ruler "behave". The velocity of a beam of light could be interpreted as having two aspects: (i) the velocity measured by a clock and a ruler before time dilation and/or Lorentz contraction, (ii) the velocity measured by a clock and a ruler after time dilation and/or Lorentz contraction; i.e., there are two different ways of interpreting the same velocity of the beam of light.

However, a close scrutiny of the Special Theory of Relativity should make one realise that it is a "theory of illusions". Concepts in the Theory such as simultaneity, i.e., what events appear simultaneous to one observer at one inertial frame might not appear simultaneous to another observer at another inertial frame, length contraction and time-dilation wherein an observer on the ground level (stationary frame) and an observer on a moving vehicle traveling at close to the velocity of light (moving frame) see one another's rulers and clocks contract in length and slow down respectively by the same degree, and, a person on a moving vehicle (moving frame) regarding himself as stationary and regarding the person on the ground level (stationary frame) as mobile, seem to be, honestly speaking, artificial. All these parties are each right in thinking the way they do, according to Special Relativity. That is, all of them are right in their thinking and no one is wrong - there is more than one reality. The invariance of the velocity of light might indeed be a true phenomenon but the explanation of this phenomenon a la the Special Theory of Relativity seems quite far-fetched. Isn't it so? For example, could we say that the law-abiding citizen is doing the correct thing by abiding by the law and the person who breaks the law, e.g., by robbing or murdering, is also doing the correct thing by breaking the law? Could we say that the person who describes a red object as red in colour and the person, perhaps colour-blind, who describes the red object as green in colour are both right? Is more than one standard or reality acceptable? Isn't all this incredible? Special Relativity appears to "preach" something like that.

The Special Theory of Relativity by postulating that the velocity of light is invariant due to clocks slowing down (time dilation) and length contraction (Lorentz contraction), each at the same rate as the other, evidently implies that without these two phenomena the velocity of light would not appear invariant; all this actually amounts to manipulating data to make the velocity of light appear invariant. Thus the invariance of the velocity of light, as is described by the Special Theory of Relativity, is evidently just an illusion.

We ask an important question: Couldn't the velocity of light be invariant without involving time dilation and Lorentz contraction? This invariance would then be indisputable.

The invariance of the velocity of light might seem as mysterious, puzzling and inexplicable as "quantum weirdness".

4 POSSIBILITY OF EXCEEDING SPEED OF LIGHT

This chapter proposes some ways to detect or confirm the existence of faster-than-light particles or tachyons which have aroused much interest and whose existence is still very much a mystery - detection is likely to be very difficult, perhaps impossible.

It is held that no object could travel faster than light for at the speed of light the object's mass would be infinitely great and the object would thus be unable to accelerate. On the other hand, an object without mass, or even with negative mass, probably a quantum particle which is in some ways similar to a photon might be able to travel faster than light, the photon being the quantum particle of light which is massless and always in motion. Such a quantum particle or quantum particles might be found in future. In any case, the tachyon, which is a theoretical particle that travels faster than light, is deemed by many to exist.

There is so far no proof of faster-than-light travel by quantum particles. Special Relativity posits that the speed of light, which is regarded as both a particle and a wave, at 186,000 miles per second is the ultimate speed for any particle, as per the following equation which indicates that no moving object could travel faster than light:-

$$v = (x + y) \div (1 + xy/z^2)$$

where x = speed of moving train, y = speed of beam of light sent from the back of the moving train to the front of the moving train with respect to the moving train, which is the moving frame - in other words, the speed of the beam of light which is sent from the back of the moving train to the front of the moving train is gauged from the moving train, which is the moving frame, v = speed of beam of light sent from the back of the moving train to the front of the moving train with respect to the ground level, which is the stationary frame - in other words, the speed of the beam of light sent from the back of the moving train to the front of the moving train is gauged from the ground level, which is the stationary frame, z = speed of light = 186,000 miles per second; if we let $x = y = z$, then:-

$$v = (z + z) \div (1 + z.z/z^2) = 2z/2 = z! \text{ (Not } 2z!\text{)}$$

The above equation may cleverly describe the speed of light as the ultimate speed for any quantum particle. However, the existence of the tachyon should not be ruled out despite the above-mentioned postulate of Special Relativity.

The photon, which is the quantum particle of light, is quite similar to the other quantum particles such as the electron, positron, proton and neutron, etc. The Michelson-Morley experiment, and other experiments, had not found any proof of the luminiferous ether as the medium of transmission for light particles. Evidently, from this "null" result the conclusion had been reached that the speed of light is constant and there is nothing which could travel faster than light. However, quantum particles have been found to have the capability of "teleportation", i.e., send information from one location to another location in space instantaneously. This is a weird and incomprehensible phenomenon, which indicates that faster-than-light travel by quantum particles is a possibility.

Special Relativity posits that the clock slows down and the length of the measuring rod contracts in the direction of motion at close to the speed of light, while at the speed of light the clock's mechanism and time are at a standstill and the measuring rod's length is nil. The slowing down of the clock (for which there had been experimental proof) and therefore time is caused by the intense gravitational field due to travel at close to the speed of light. However, this evidently is not sufficient as an explanation for the contraction of the measuring rod's length in the direction of travel at close to the speed of light, which is only an inference. There are thus serious implications for the gauging of the speed of light and similar particles.

There have also been some speculations with regards to the variable speed of light (VSL). One theory, for instance, states that the speed of light changes with the various stages of the evolution of the universe, exceeding 186,000 miles per second at certain times.

The speed of light is obtained by dividing the distance traveled by the beam of light by the time which the beam of light takes to travel this distance.

In order for the speed of light to stay constant, on approaching the speed of light the clock has to slow down to the same degree as the contraction in the measuring rod's length, which is in accordance with Special Relativity. Below are the possible outcomes of the slowing down of the clock and the contraction of the measuring rod's length, when the speed of light is not constant, i.e., it is variable:-

$$T^p > M^p \rightarrow O^v \quad (1)$$
$$T^p < M^p \rightarrow P^v \quad (2)$$
$$T^p = M^p \rightarrow Q^v \quad (3)$$

where T^p is the percentage of slowing down of the clock, M^p is the percentage of contraction in the measuring rod's length, O^v is the increase in the speed of light, i.e., exceed 186,000 miles per second, P^v is the decrease in the speed of light, i.e., go below 186,000 miles per second, Q^v is the speed of light, 186,000 miles per second

An object's mass or inertia prevents the object from accelerating beyond the speed of light, i.e., 186,000 miles per second. Since light particles, i.e., photons, are without mass or inertia, there is in theory nothing to prevent them or other particles with no mass or inertia, or even with negative mass, from traveling more than 186,000 miles per second. The detection of these faster-than-light particles would be difficult. The reasons for this difficulty are given below.

In order for tachyons, or, faster-than-light particles, to be detected, some obstacles have to be overcome. When seeing or detecting, light is needed to bring images to our eyes. Should a particle, for instance, the tachyon, move at a speed greater than that of light whereby light cannot catch up with it, this faster-than-light particle would evidently not be detected by the light. If such a faster-than-light particle exists, we would thus not be able to detect it. How could this problem be solved?

Tachyons have long been the subject of investigation. There have been speculations about faster-than-light travel; a number of scientists believe tachyons exist though this is yet to be proved. For instance, a reputable scientist has posited that when the universe first formed the speed of light had been greater than it presently is. Photons, the light particles, have no mass, while tachyons may in theory have negative mass. Thus tachyons and light appear very dissimilar. Everything about tachyons is now just postulation and not scientific fact, unless their existence could be confirmed by experiment. However, this experimental confirmation seems to be very difficult to obtain.

The speed of light is constant; nothing can exceed this speed, according to Special Relativity. Experiment had also confirmed this. Despite this, the existence of tachyons or faster-than-light particles should not be ruled out, no matter how low the chances are.

Light is necessary for detecting tachyons for our eyes cannot see in the darkness. Light would reflect images into our eyes so that the images would be seen. If light particles are slower than the faster-than-light particle and cannot catch up with it, light cannot reflect its image back to our eyes or a sensitive detector. Einstein had posited that a faster-than-light particle has negative mass and length, which implies that the faster-than-light particle has suddenly changed its travel route to the opposite direction - this could be interpreted as traveling backwards in the time dimension, which appears absurd, weird. It probably coerced Einstein to postulate that the speed of light is the maximum speed any particle could attain.

Special Relativity also postulates the slowing down of time when the speed of light is approached. In order for the speed of light to stay invariable, the slowing down of time and the shortening of distance traveled or length contraction must occur at the same rate, as is described earlier. Einstein had apparently been greatly mystified by the invariance of the speed of light. It is actually not far-fetched for a particle to exceed the speed of light when the fact is that in quantum entanglement particles could inter-act instantaneously and in electronics a TV signal sent out from a broadcasting station could be received instantaneously in all the homes tuned to it at the same moment. Though tachyons are now only a speculation without any proof of existence, they should not be viewed as an absurd thought with no possibility of existence.

Relativity posits that a tachyon travels back in time into the past. But there is a possibility of the tachyon traveling forward in time. In the case of quantum entanglement a particle sends information to another particle instantaneously, i.e., with faster-than-light speed. This information could have been sent through a carrier wave comprising of tachyons that travels faster than light.

The detection of tachyons is evidently a very difficult task. Quantum entanglement, which has caused puzzlement to many scientists including Einstein, is perhaps a sign of the presence of tachyons, which is yet to be proved or confirmed by experiment. Tachyons may either exist or do not exist, or, they may be moving too fast to be seen, or, they may have annihilated or negated the light particles when bumping into the latter. There is still a lot of mystery surrounding the tachyons.

Though there is still no evidence that negative mass particles exist, a tachyon is possibly a negative mass particle. A negative mass particle could annihilate or negate a non-negative mass particle such as a light particle or photon. If negative mass particles or negative mass tachyons do not exist, i.e., only positive mass tachyons exist, there is evidently a greater possibility of detecting tachyons, as light particles needed for their detection would not be annihilated or negated by these negative mass tachyons. However, if the light particles were indeed annihilated or negated it might be construed that negative mass tachyons exist, which explains why the light particles were annihilated or negated. This would be circumstantial proof of the existence of negative mass tachyons.

It is probably not possible to detect a negative mass tachyon directly in the usual way with light particles and the only way is probably to use an indirect method of proof, i.e., finding some kind of circumstantial proof. It would evidently be very hard to ascertain that a tachyon is a negative mass particle and not a positive mass particle. One way is to just assume that the tachyon is negative mass, or, positive mass, and make deductions from there and see whether either assumption leads to contradictions or problems. Inference may be the only recourse and may thus suffice for proving that the negative mass tachyon exists as physical proof may be impossible to obtain due to the reason given above. This method has some promise.

There is a possibility, in theory at least, of detecting the speed of a particle exceeding the speed of light without using light as a detector. Like the case of timing a runner from the starting point to the finishing line, the timing of the route taken by the tachyon from creation to destination might be possible, if a tachyon could be created in the laboratory. The tachyon's creation, for instance, could activate a timer which stops when the tachyon hits an obstruction - the destination. The timer could be synchronized to carry out this function. The time gauged in this way is the time the tachyon takes to travel from its creation (by some laboratory equipment) to its obstruction - destination. This is the most direct way of detecting and gauging the speed of a tachyon.

Researchers at CERN had mistakenly thought that faster-than-light neutrinos were discovered, in 2012. This had been found later to be untrue as it had been due to a defective experiment which had a cabling problem. The discovery of tachyons would indeed be a very important scientific achievement.

5 CONCLUSION

The Special Theory of Relativity postulates that on approaching the speed of light a person's brain, as well as bodily, functions slow down, which partly explains why the speed of light would remain unchanged to him under all circumstances; in other words, he experiences the slowing down of time on approaching the speed of light. It implies that the person's mind could travel backward and forward in the space-time continuum. The mind is evidently closely linked to the "time" dimension of the space-time continuum, while the three coordinates of length, breadth and height make up the "space" dimension of the space-time continuum. We could equate the mind and time as follows:-

$$C = S^t$$

where C represents consciousness and S^t represents sense of the passing of time

Thus, there ought to be a Theory of Consciousness allied to the Special Theory of Relativity, which would contribute to an important understanding of the workings of nature.

BIBLIOGRAPHY

[1] A. Beiser, Concepts of Modern Physics (Fifth Edition, McGraw-Hill, 1995)

[2] Andrew L. Bender, SlipString Drive: String Theory, Gravity, and "Faster Than Light" Travel (iUniverse, 2006/2007).

[3] David Bohm, The Special Theory of Relativity (Routledge, 1996)

[4] Albert Einstein, "On the Electrodynamics of Moving Bodies", Annelen der Physik 17, 891 (1905).

[5] Albert Einstein, The Meaning of Relativity (Princeton, 2005).

[6] Joao Magueijo, Faster Than the Speed of Light: The Story of a Scientific Speculation (Perseus, 2003).

[7] Paul Tipler, Ralph Llewellyn, Modern Physics (4th. Ed., W. H. Freeman Company, 2002)

www.ingramcontent.com/pod-product-compliance
Lightning Source LLC
Chambersburg PA
CBHW081018170526
45158CB00010B/3087